Ecological Tourism in the Republic of Kazakhstan

Katima Iskakova · Sairan Bayandinova ·
Zhannat Aliyeva · Aliya Aktymbayeva ·
Ruslan Baiburiyev

Ecological Tourism in the Republic of Kazakhstan

Springer

Katima Iskakova
Al-Farabi Kazakh National University
Almaty, Kazakhstan

Sairan Bayandinova
Al-Farabi Kazakh National University
Almaty, Kazakhstan

Zhannat Aliyeva
Al-Farabi Kazakh National University
Almaty, Kazakhstan

Aliya Aktymbayeva
Al-Farabi Kazakh National University
Almaty, Kazakhstan

Ruslan Baiburiyev
Al-Farabi Kazakh National University
Almaty, Kazakhstan

ISBN 978-3-030-77461-5 ISBN 978-3-030-77462-2 (eBook)
https://doi.org/10.1007/978-3-030-77462-2

This Springer imprint is published by the registered company Springer Nature Switzerland AG
The registered company address is: Gewerbestrasse 11, 6330 Cham, Switzerland

Preface

Ecological tourism is currently one of the most rapidly developing areas of the global tourism industry. It is aimed at addressing many important issues that are of paramount importance today, including the prevention of a gradual deterioration of environmental situation throughout the world, the preservation of biological diversity concerning recreational natural areas through the rational use of recreational resources, the rehabilitation of public health and, in general, the sustainability of tourism itself.

Monograph was compiled in accordance with the country's development priorities set forth in documents such as «Basic principles and general approaches for the tourism industry development in the Republic of Kazakhstan»; decree of the Government of the Republic of Kazakhstan dated from June 30, 2017, No. 406 «On approval of the concept for tourism industry development of the Republic of Kazakhstan until 2023»; decree of the Government of the Republic of Kazakhstan dated from 23.12.2013 No. 1385 «On the concept of Kazakhstan becoming one of the 30 most developed states in the world»; «Development of the service sector» in paragraph 3, «Vision of Kazakhstan in 2050» (decree of the President of the Republic of Kazakhstan dated from 17.01.2014 No. 732); paragraph 4 «Strategic directions for the development of regions» in the «Forecast scheme for territorial and spatial development of country until 2020» decree of the President of the Republic of Kazakhstan dated from 21.07.2011 No. 118, which justifies the priority of ecological tourism in the tourism structure of the Republic of Kazakhstan.

These documents outline the need for a competitive tourism industry to ensure stable income growth and employment in hospitality by increasing inbound and domestic tourism; ensuring transport accessibility of tourist destinations and facilities; improving the quality and accessibility of tourism products and services; creation of a favourable tourist climate in the country; increasing the investment attractiveness of tourism industry; improving the marketing and promotion of national tourism brand; and improvement of the tourism industry management system.

Achievement of the objectives set out in these documents is possible only with systematic studies of ecological tourism current state in the country regions, which underlies many other types of tourism activities. There is a need for comprehensive assessment of tourism and recreational potential and, more importantly, a scientifically sound selection concerning promising types of tourism activities.

In this regard, the topic of research on ecological tourism is of particular relevance and priority.

The aim of this monograph is to assess the tourism and recreational potential for the purposes of scientifically based territorial organization concerning ecological tourism in the Republic of Kazakhstan.

Main sources of information were primarily statistical, reference, stock, cartographic and literary materials on the natural and cultural–historical resources, and tourism infrastructure of Kazakhstan.

In the process of performing the work, a set of methods and approaches was used: cartographic, mathematical–statistical, comparative–geographical, historical, systemic, spatial network, rationing and point estimates.

Conclusions of assessment on tourism and recreational potential for the development of domestic and ecological tourism contribute to an integrated approach in territorial organization of promising special tourism types. Resulting cartographic material visualizes in the geospatial potential of Kazakhstan regions for more targeted spending of financial resources allocated for the domestic tourism development, which are of value to business structures and stakeholders. Thus, the presented monograph is also relevant from a practical point of view.

Monograph analyses theoretical and methodological foundations of ecological tourism development, taking into account the essence, content, factors and models of ecological tourism development.

Tourism and recreational potential of the Republic of Kazakhstan are huge. Ecological tourism development is especially important for the country, since this direction of tourism is a favourable environment for the functioning of small businesses. Authors conducted studies to assess the tourist and recreational potential of ecological tourism in the context of Kazakhstan regions. Analysis of the current state is given, and the problems of promising ecological tourism development are described. In addition, the role of specially protected areas in ecological tourism development is shown and a list of organizations capable of creating tourist products in the environmental direction is given. Main conditions necessary for ecological tourism development in protecting the natural and rural areas have been determined.

Moreover, in conditions of world economy instability, increasing environmental risks, more efforts are needed to find and implement mechanisms to reduce the ecological burden on the environment, alternative and «sparing» methods of spending irreparable natural resources. This approach is being pursued by many developed countries considering tourism as an alternative. At the same time, one of the effective measures to preserve and restore the natural environment is to intensify ecological tourism and environmental education in the country.

Almaty, Kazakhstan

Katima Iskakova
Sairan Bayandinova
Zhannat Aliyeva
Aliya Aktymbayeva
Ruslan Baiburiyev

Contents

1 Scientific and Methodological Basis of the Ecological Tourism Development

1.1 Essence and Content of Ecological Tourism

Among the main prerequisites for ecotourism emergence, the leading role is played by increasing anthropogenic load on the natural and cultural-historical tourism resources. This load increases in direct proportion to the growth rate of tourist visits. Considering the tourism development indicators predicted by the World Tourism Organization in the XXIst century, betterment in contradictions, which concerns satisfying tourism demand and rational use of tourism resources becomes apparent.

Stemming from the environment breakdown in each field of world economy, an explored is underway for alternative ways of using the natural resources. Their main characteristic is the reduction of environmental damage agreeing sustainable developing perception.

Fundamental changes are taking place in the tourism field. Now it is becoming clear that with each year an increasing tourist migration poses is a threat to the natural environment. This threat is manifested in destruction of landscapes as a result of excessive people's concentration in well-known rest places, in the accumulation of rubbish and waste, in the noise created by speakers and transport means. Motorization of tourism played a large role in air and water pollution. The development of mass tourism without regard to environmental requirements led to the gradual degradation of the natural complexes. One of the reasons for recreational areas' degradation is the lag in tourist offer from tourist demand, i.e. insufficient development of recreational industry concerning the number and variety of objects, which leads to an excessive concentration of holidays in familiar and long-known areas. In this regard, there are many countries which carried out a planned territorial organization of tourist traffic, and it leads to the development of this tourism type, which ensures equilibrium the natural complexes. This tourism type has received the general name "ecological tourism".

Currently, ecological tourism is widely used in modern activities as one of the main aspects concerning environmental education of people. This is one of the new areas of modern tourism, which in recent years has been widely developed

K. Iskakova et al., *Ecological Tourism in the Republic of Kazakhstan*,
https://doi.org/10.1007/978-3-030-77462-2_1

throughout the world. Until recently, environmental tourism in Kazakhstan was not widely used, but in the last decade, much more attention has been paid to this issue in our country.

The term "ecotourism" appeared at simultaneously with the term "sustainable development". Ecotourism is travel to the natural places not contaminated and untouched by human activities to adore beautiful landscapes, wild plants and animals, as well as additional existing cultural characteristics. At the same duration, it does not spoil the integrity of ecosystems and originates such economic conditions under which environmental protection becomes beneficial for local population.

Thus, ecological tourism pursues the goal of preserving the natural environment in the natural state and at the same time brings a certain profit, satisfying the interests of residents.

Peculiarity of the modern stage in tourism development can be said to be revealed through the ecological tourism concept.

The term "ecotourism" appeared at simultaneously with the term "sustainable development". As noted above, there is a definition given by the International Commission on Environment and Development, "sustainable development" is a betterment in which the issues of present day are resolved without compromising the interests of future generations.

However, the exact criteria for determining truly stable mechanisms of the relationship between man and nature have not yet been found. Therefore, it is necessary to be careful when it comes to any activity affecting the environment, study its consequences and quickly identify erroneous actions.

Growing popularity of ecotourism is determined by a combination of increased demand for related services and increased need for nature conservation. Considering the need to preserve ecosystems, ecotourism presents unique opportunities for maintaining the ecological balance of the planet.

It should be noted that in Kazakhstan, few people know what is ecotourism in general, how to develop it and what income it can bring. Meanwhile, in the modern era of progressive environment degradation, the main goal of ecological tourism should be to achieve a balance of recreational use concerning the natural resources (Lindberg 1991).

To understand the scope of this tourism industry, it is first necessary to provide some clarifications regarding its definition. It should be noted that the definitions of eco-tourism are still being developed, however, in general, they include recommendations for travelling to areas with a well-preserved the natural environment and creating conditions that improve the well-being of local residents. There are various definitions of ecotourism. According to some reports, the term "ecotourism" was determined in 1983 by Mexican ecologist Hector Ceballos-Lascurain. Later, in 1991, the International Ecotourism Society based in the United States proposed a slightly modified definition of ecotourism, in which it does not violate the integrity of ecosystems and creates such economic conditions that environmental protection becomes beneficial for the local population. In our opinion, this definition more fully features the essence of ecotourism.

Overall, there are two main approaches to defining the ecological tourism sector. In the first case, ecotourism is referred to as tourism, the main object of which is "wildlife". Ecotourism as a tourist activity was formed in the United States of America in the 1980s in line withe development of tourism for the naturalists in the most remote and vulnerable regions of the world. Its most frequently cited definition is "Ecotourism is tourism within territories not actually affected by the existence of mankind. It should promote the conservation of nature and the well-being of local population" (Turizm I dovkillya: Serija «Pitannya ta vidpovidi» 1999).

To give a definition that more fully characterizes the essence of ecotourism, it is necessary to familiarize yourself with the history of tourism development through the lens of data at a certain moment of their concepts (Table 1.1).

In our opinion, these terms do not fully disclose the concept of ecotourism, since in different years and in different countries the role of ecotourism is various. Some terms either do not fully reveal the essence of ecological tourism, or are very detailed, which makes the concept itself narrower. Thus, the term "ecotourism" is used very frequently, but its use is not always justified, since the definition boundaries are rather blurred. However, the answer to the question about the real nature of ecotourism was given in due time by the Quebec Declaration on Ecotourism, which distinguishes this concept from the more general concept of "sustainable tourism" according to a number of criteria that an ecotourism product must meet.

Among the basic principles of ecological tourism, researchers name:

- minimizing negative impacts on the nature and culture of territories visited;
- participation of local population in tourism development and their receipt of financial and other benefits from tourism activities;
- improving the environmental culture of participants in ecotourism;
- favourable promoting concerning the conservation of the natural and cultural heritage of recreational areas.

In our opinion, paragraphs 1 and 4 can be just attributed to the basic principles of ecological tourism, but paragraphs 2 and 3 still relate more to the consequences that play a role in the levels of ecological tourism development. If the local population is not actively involved in ecotourism development and culture of its participants is not at a high level, then the tourism development level in this destination will be low. On the other hand, as David Weaver noted, ecotourism does not exist in a separate context from culture.

Above terms cannot be synonymous and equated with each other for some reasons: the concept of sustainable tourism provides for tourism industry integrity and covers both agricultural and urbanized areas and the natural landscapes, including the idea of cultural and architectural heritage, and calls for changes in behaviour both within their country and in the country that is visited.

Table 1.1 Definitions of ecotourism [compiled by the author (authors)]

Year	Authors	Definitions
1	2	3
1983	Hector Ceballos-Lascurain	A tourism industry that includes travels to relatively untouched or unpolluted the natural areas with inherent objects to delight, explore, enjoy the beautiful landscapes and its wild plants and animals, as well as other existing cultural features
1988	K. Miller	Viable tourism is environmentally sensitive, including the study of the natural and cultural environment and aims to improve the situation in this environment
1990	International Ecotourism Organization	Responsible journey to the natural zones, areas that preserve the environment and support the well-being of residents
1991	International Ecotourism Society, United States	Travelling to areas with a relatively well-preserved the natural environment to get an idea of the natural and cultural-ethnographic features of the area
1992	Commonwealth Department of Tourism, Australian National Ecotourism Strategy, Canberra	Nature-based tourism that involves education and interpretation of the natural environment and is managed to be ecologically sustainable
1993	International Ecotourism Society, United States	Responsible trip to the natural areas that preserves the environment and improves well-being of residents
1993	K. Lindberg	Sustainable and nature-oriented tourism and recreation
1993	Valentine P.S.	Tourism, which takes place in relatively undisturbed the natural territories, does not lead to the destruction of the natural environment, its quality deterioration, makes a direct contribution to the protection and management of the used the natural territories, moreover it is subject to adequate and competent management
1994	World Wildlife Fund	Tourism, including trips to places with relatively untouched nature, to get an idea of the natural, cultural and ethnographic features of the area, which does not violate the integrity of ecosystems and creates economic conditions in which protection of nature and the natural resources becomes beneficial for local population
1994	World Tourism Organization	Controlled form of nature tourism, in which tourists make hiking or boat trips in the natural areas, accompanied by local guides who provide explanations about the local flora, fauna, ecology. Ecotourism may include visits to villages and farms

(continued)

Table 1.1 (continued)

Year	Authors	Definitions
1	2	3
1995	Society of Ecotourism in Germany	A set of tourism forms, in which a conscious attempt is made to minimize the negative impact on the environment, to assist in financing protected the natural areas, to create sources of income for the local population
1997	Marcus Endicott	Travel with low anthropogenic environmental impact or tours to nature
1997	G. S. Guzhin, M. Yu. Belikov and E. V. Klimenok	Journey, which is based on concern for the environment and comes to for the organization of trips with a limited number of participants to the natural zones with possible visits to places of cultural interest on implementation of various projects for protection and rational use of the natural resources
1999	Tourist Terminology Dictionary	Combining travel with environmentally friendly attitude to nature, allowing to combine joy of familiarity with new landscapes, study of flora and fauna samples with the possibility to contribute their protection. The main resource of ecotourism is protected areas—reserves, national and nature parks, unique and typical landscapes. Ecotourism economically contributes to nature conservation: protected areas attract tourists, for which jobs are created; the natural resources become recreational, and their use is based on compliance with environmental laws
1999	Canada's Advisory Council on the Environment	A type of nature-related tourism that contributes to conservation of ecosystems while respecting the interests of local people
1999	Fennel D.	An ecologically sustainable form of nature tourism, focused primarily on living in the wild and knowing it, organized in accordance with ethical standards in such a way as to minimize environmental impact, consumption and costs, and oriented to the numbered level (in terms of control, benefits and scope)
2001	David Weaver	Ecotourism is a form of tourism that fosters learning experiences and appreciation of the natural environment, or some component thereof, within its associated cultural context
2004	Yu. L. Mazurov	Active form of recreation based on sustainable use of the natural goods
2005	Swiss School of Ecotounsm	Silent or quiet tourism, which should provide physical and spiritual recreation to people in direct contact with the landscape, considering the long-term interests of residents

(continued)

Table 1.1 (continued)

Year	Authors	Definitions
1	2	3
2008	International Society for the Survival	Ecotourism promotes the interests' primacy of local people in tourism development of territory, protects the local flora and fauna and provides economic incentives for local people to preserve the environment
2015	The US Ecotourism Society	Any kind of tourism and recreation in nature that does not damage the natural complexes contributes to the protection of nature and improvement of the local population well-being
2015	IUCN	Journey with responsibility to the environment in relatively undisturbed the natural areas for purpose of studying and enjoying nature and cultural attractions, which contributes to nature protection, has a mild impact on the environment, ensures active social and economic participation of residents and benefits from these activities
2015	N. N. Mamayeva	A form of the natural tourism, which is usually carried out in relatively intact corners of nature for ecological education, familiarization with traditional lifestyle, crafts, folklore traditions and historical past. Main conditions for such tourism are the careful use of the natural resources, preservation of the natural and cultural complex and support of the local, mainly traditional economy

Therefore, ecotourism definition, in the opinion of authors, should be as follows: "Ecological tourism is a type of tourism related to its needs, contributing to the environment preservation and based on tourist demand developing within the cultural context".

As was noted above, the definition "ecotourism" was introduced into everyday life quite recently, although as a certain kind of recreation it arose many years ago. At the beginning of the nineteenth century, Europeans travelled to Africa and India to see wild life (safari). The first organized American ornithological travels took place in the first half of the twentieth century, and the first ornithological trips in Europe started in 1961 (Naibolee seryoznye sosialnye problemy, volnuyushie lyudei 1998). Present day, ecotourism has raised to one level or another in almost each country.

Along with the establishment and separation of ecological tourism into an independent type of tourism activity, a process of rapprochement with other types of tourism is noted. The desire for nature is also present in the mass tourist movement. Swimming and beach holidays, cruises, business trips often include visits to the natural or national-ethnographic parks, as well as nature reserves, wildlife sanctuaries. The borders of ecotourism itself are becoming more blurred and difficult to describe. However, in our opinion, there are two main approaches in distinguishing the environmental sector from the totality of tourism-related phenomena.

The first of them is based on presence of wildlife as the main tourism object. However, most authors note the difficulty of dividing the border between nature and traditional culture and, along with nature, include the latter in ecotourism objects. However, even in this situation, the capacity of this concept is not fully exhausted. Tourism is widespread for the purpose of outdoor recreation in areas significantly altered by humans. This is especially true for developed countries, where an increasing demand for ecotourism services is satisfied both by visiting the territories of historical and cultural heritage, and by relaxing in the nature countryside. Despite this, such tourism is often categorized as ecological, and its importance for the protection and restoration of nature, traditions and greening of economic development becomes conclusive in any area. A demonstrated of such activity is agro-tourism.

The second approach highlights the position that ecotourism is tourism, which, being characterized by a lower the natural resource intensity relative to traditional tourism, does not lead to degradation of the used the natural and other resources. In this case, the term "environmental" is equated to the term "environmentally friendly" or "sustainable", and tourism objects can be both the natural and, to one degree or another, artificial (Panov 1998).

Thus, it is quite difficult to draw clear boundaries for ecotourism. Nature-oriented (by objects) tourism can also develop with complete neglect of the sustainable use of resources pursuing high profits, or without proper management and planning. This experience discredits the term "ecotourism". On the contrary, environmental friendliness can be characterized not only exclusively by nature-oriented tourism. Moreover, to assess the environmental impact in conditions of so-called "vertical integration" concerning main links of the tourist technological chain (transport, consumer services and actual services for acquaintance with tourism objects), it is necessary to consider all three links. For example, transport services are difficult to attribute to nature-oriented tourism, even if they serve this type of tourism. In addition, it is almost impossible to clearly determine the degree of resource degradation and possible losses during the tourism development. And tourism, traditionally considered sustainable, can lead to negative unpredictable consequences eventually. Therefore, it is necessary to combine the two approaches mentioned above.

Sustainability of tourism is achieved when part of the benefits derived from it goes to the renewal (protection) of resources and elimination of the negative impact concerning production process on environment and quality of resources. Otherwise, depletion or degradation of resource is inevitable. Here, environment and resources are understood in their broad sense and include not only the natural, but also anthropogenic components.

An integral characteristic of ecotourism is environmental education. It includes two aspects:

1. Informational—obtaining knowledge about the nature and culture of territory, environmental patterns, etc.;
2. Ethical—the formation of ecological mentality—respect and understanding of nature and traditional cultures, awareness of the need to preserve them.

Thus, the objects of ecological tourism can be various manifestations of animate and inanimate nature, traditional cultures and cultural heritage, inseparable from the natural environment. Main goals pursued by ecotourists are the objects' cognition, while outdoor recreation and sports tourism play an important role. This allows you to highlight the cognitive and associated with outdoor activity elements of ecotourism.

Ecotourism is one of the new directions of modern tourism, which has received wide development in recent decades throughout the world. This is evidenced by the involvement of well-known international organizations in ecotourism development activities, such as the World Wildlife Fund (WWF), International Union for Conservation of Nature and The natural Resources (IUCN), The International Ecotourism Society, Ecotourism Associations, Travel Green Guide, Eco Travel 360°. Among other things, there are many congresses, seminars, symposiums are held on the ecotourism development at the international and regional levels, furthermore the number of companies is increasing that prefer ecological tourism, among all types of tourism.

The growing popularity of ecotourism in the world is explained not only by environment deterioration, but also due to the acceleration of changes in human environment, and growth of psychological stress. For example, the World Tourism Organization declared 2020 as the year of rural and ecological tourism. In 2020, UNWTO will conduct all its events not in capitals, but in provincial cities and regions.

What is the difference between ecotourism and conventional tourism?

Firstly, ecotourism is a visit to territories with well-preserved the natural environment. Such territories include nature reserves, national nature parks, wildlife sanctuaries and other protected the natural areas.

Secondly, ecotourism is the most "environmentally friendly" nature user. The technology of nature exploitation by ecotourists is such that they "extract" most of the useful resources visually (sightseeing, admiring landscapes, observing the flora and fauna), as well as by ear (birds singing, murmur of stream, noise of foliage), smell (smells of forests and meadows) and touch (freshness of river water). For this reason, ecological tourism has become the only type of the natural resources use within the majority of specially protected the natural areas (SPNA) of the world.

Thirdly, ecotourism strictly regulates the tourists' behaviour in protected areas, while defining rather strict rules of behaviour (tourist should use a special path accompanied by guide, do not pick plants, make noise, etc.). Compliance with them is a fundamental condition for successful development of ecological tourism.

Fourth, ecotourism provides an opportunity for residents to improve their well-being, working as support staff, as well as through the development of traditional national arts and crafts that contribute to the revival of national customs and traditions. All this will bring a certain income to the population, and contribute to its socio-economic development (Shtyurmer 1978).

Ecotourism is understood as a spatial phenomenon. Its development helps to improve the infrastructure of settlements located on ecological routes territory, and increase employment, while reducing its migration.

Therefore, ecotourism contributes to environmental education and upbringing of the population. Peculiarity of the process concerning environmental training and education in ecotourism can be understood as it is built based on not didactic-edifying, but easy assimilation of information, behavioural norms in the natural environment (Ivanov and Chizhova 2003).

As noted above, ecotourism is very closely related to the protection of rare and endangered species considering plants and animals, since many of them are exotic and become objects of ecotourism.

It should be recognized that ecological tourism has broad prospects for economic development. The growth of tourist flows contributes to economic integration of world economy—a process that allows more efficient use of the planet resources. Currently, income from ecotourism continues to grow rapidly (Alexandrova 1996).

Therefore, in addition to the fact that ecological tourism contributes on the formation of the right relationship between man and nature, it serves as a considerable source of foreign currency for countries hosting tourists.

It should be noted that ecological tourism, in addition to the natural objects, includes cultural and architectural heritage objects in the circle of tourist interests and contributes to their conservation, as well as to the interests' satisfaction of population living in visited territories.

Ecotourism is often seen as an alternative to mass tourism, and now it is one of the most promising and dynamic forms of tourism activity, developing to one degree or another all over the world. In some countries, ecotourism is not only the largest tourism sector, but also the most important sector of economy, in others—a means of nature conservation. In the transition to sustainable development, ecotourism, in comparison with other types of tourism, can ensure economic and environmental balance.

Due to the fact that the economic development level and state of the environment concerning macro-regions of the world are not the same, developed and developing countries are characterized by their characteristics of transition to sustainable development.

In developing countries, the first condition for sustainability is economic development. Significant areas of the natural territories here have been preserved in their natural state or have experienced a small anthropogenic impact than in developed countries. Often irreversible changes are occurring in territories covered by extensive economic activity, mainly ineffective agriculture, which is the only livelihood for an ever-growing population. Only the economic development of these countries will lead to decrease in anthropogenic pressure on the environment and will help preserve the natural territories important for maintaining the ecological balance of the planet, and traditional cultures that are world heritages.

Economic development is impossible without an increase in human and social capital, and growth of which is one of the conditions for liberation of developing countries from economic dependence.

Developed countries have achieved a high level of human development, mainly due to changes in the natural environment. In these countries, a small proportion of untouched territories has remained. The change in the natural environment, first, affects the quality of life, which is smoothed, however, by the growth of economic prosperity. Therefore, landscapes restoration has great importance, and, as far as possible, their maintenance. Thus, developing countries have predominantly the natural capital, but its use to increase wealth is ineffective. Developed countries, on the other hand, have significant human and social capital, and the lack of the natural environment is compensated by active ecotourism in developing countries. The natural environment in developed countries is less stable than in developing ones. Social, by contrast, is more stable in developed countries.

In September 2015, UN member states approved the 2030 Agenda for Sustainable Development and 17 Sustainable Development Goals. The Sustainable Development Goals constitute a comprehensive agenda for equitable, socially inclusive and environmentally sustainable development. They form a common basis for solving the most pressing economic, social and environmental problems of the current generation, including regarding the roles in our society must take on to achieve sustainable development (Vsemirnaya konferensiya Turizm i energiya budushego—Sokrashenie vybrosov CO_2 2017).

Countries with transit economies, which include the states of Central and Eastern Europe, as well as countries entering the Commonwealth of Independent States (CIS), are experiencing simultaneously problems typical of both developed and developing countries (Volsky 1998). Thus, in European countries with transit economies, the significantly altered the natural environment is combined with social problems. In the CIS countries, and especially in Russia, Kazakhstan, large areas of the natural territories with a low degree of anthropogenic variation have been preserved, and the instability of socio-economic development determines an excessive burden on nature. Unfortunately, mainly these countries have focused their main efforts on achieving sustainability of social and economic development harming the natural sustainability. In this, countries with transit economies are similar to developing ones.

At present, when the Republic of Kazakhstan is at the beginning of solving the problem which creates a modern highly efficient tourism industry, consideration of ecological tourism development is very relevant in this situation. Many territories of our republic in the case of their tourist use can give a much higher income than when using them for agriculture and industry.

Despite the popularity of ecotourism around the world, few people in our country know what ecotourism in general is, how to develop it and what income it can bring. Meanwhile, in the modern century of advancing degradation regarding the natural environment, the leading occupation of ecological tourism should be to achieve the recreational balance use regarding the natural wealth.

A clear understanding of ecotourism essence, its functions and development paradigm will help to save the natural complexes, reduce the negative consequences of human recreation in nature, increase awareness and responsibility of sightseers, attract the local population to organize all-purpose appearances of tourism activity and profit through the implementation of a complex on tourist and environmental activities and tourism infrastructure development.

1.2 Role of Ecological Tourism in the Tourist System

Globalization and integration of the economic, political and information space has led to changes in the tourist space. Tourism as a systemic education is the organized set of the tourism system elements (entities, objects and means of tourism, various types of tourism entity activities, etc.) that are in certain relationships among themselves and form a certain integrity and unity.

Tourism is a complex system object with a variety of internal connections and external relations with other systems. Tourism system can be defined as a set of elements and activities that are directly or indirectly comprehended in goods production to meet the many needs of tourists, and which, in accordance with the evolution of social, economic and political relations, are always developing, becoming more complicated, and acquiring new contours (Novikov 2013a, b).

Tourist system is an open system that deals with different several exterior systems, interconnects tourist system elements, and exchanges resources and information with them.

In order to distinguish the tourism system, it is essential to know its characteristics, purposes and how its structure is organized. Among the main properties of tourism system are its openness, integrity, hierarchy, stability and synergy (Table 1.2).

Tourism system has the feature of self-development. Self-development implies mechanisms that ensure self-preservation and improvement of the system. Such mechanisms include self-organization, self-government and targeting (Table 1.3).

Purpose of the tourism system can be determined by realization of social, economic, environmental, cultural and other objectives arising from the improvement of social relations in world community.

At the heart of the tourism system is a structure consisting of elements that perform certain functions, their activities affect system properties. This structure operates through a mechanism of orderly relationships and relationships between elements.

Each of the subsystems performs relatively independent functions aimed at accomplishing the main objective of tourism system is directly or indirectly contained in benefits creation usually required not only for the tourist to meet his or her numerous demands and needs for contemporary knowledge, communication and recreation, however also for the local population. Therefore, tourism subsystems are

Table 1.2 Features of the tourism system [compiled by the author (authors) based on the materials (Pisarevskaya 2014)]

Features	Description
1	2
Integrity	During the tourism system operation, internal relationship (spatial–temporal, causal) between its elements is higher than the system strength of elements' connections to external environment
Emergent	As relationships develop between elements, the system acquires new features that are not inherent in its elements individually
Hierarchy	Tourism system can be seen as a component of some higher level (social, economic) system, and elements of the tourism system can act as systems at lower levels
Stability	Ability of the system to respond on impacts and threats of external environment (unpredictable man-made disasters, the natural disasters, political, economic and social situations that are accidental) by adapting behaviour to new conditions, preserving or changing the structure when certain elements of it disappear and other internal problems arise
Synergism	Overall functionality of the system is better than that of all its constituent elements combined, the cumulative effect exceeds simple sum of action results on all its elements

Table 1.3 Mechanisms ensuring self-preservation and improvement of the tourism system [compiled by the author (authors) based on the materials (Pisarevskaya 2014)]

Mechanisms	Description
1	2
Self-organization	Process of creating relationships and sharing functions among system elements
Self-government	Interaction of system parts, aimed at implementation of control actions with the use of reverse links and promoting self-preservation on system
Goal-setting	Formulation and selection of functional objectives, development of the system activity direction for their implementation, coordinates the objectives of its constituent subsystems

open, as they have their own indicative peculiarity of betterment and encompassing connections with independent environment.

Elements of the tourism system gather several groups of interacting topics on tourism activity.

1. Individuals (tourists) and their associations.

Subjects of the tourism system are individuals with such attributes as the desire to know new and to understand the past; need for life support, communication and experience; activity in actions, including concretized actions; ability to predict and others. Taken together, these attributes simulate the behaviour of systems to a certain extent, and in the reel the desires, aspirations and needs of individuals are shaped by the iodine influence of these systems. Since everyone is a member of many several systems, their preferences and needs are complex.

Individuals based on common interests can form clubs and other associations of tourism service consumers—public non-profit organizations and public associations, which will be elements of the tourism system, in accordance with the procedure established by legislation.

Activities of such associations can be aimed at the organization and betterment of amateur tourism, protection of tourists' rights and interests, formation of public opinion on ways of tourism progress and activities of tourism industry organizations, to solve other issues and problems.

2. Entities producing and providing goods and services and facilities of consumer value to tourists.

Including individual people to whom, together with legal entities, legislation grants the right to engage in business activities without the formation of a legal entity from the moment of state registration.

3. Intermediaries organizing tours and offering a tourism product on market. These include tour operators, travel agents, consulting organizations.
4. Non-profit organizations of entrepreneurs, employees of tourism industry, consumers of services, associations, unions and other associations promoting the growth of tourism sector.
5. Institutions involved in tourism training and scientific evolution.
6. Public authorities exercising legal regulation in the field of tourism, national, regional institutions engaged in management of tourism activities, state unitary or joint-stock enterprises.
7. International intergovernmental organizations, as well as international non-state organizations and their specialized agencies promoting tourism activities.
8. Local population living in the territory of destinations, tourism centres and in other populated places visited by tourists (Novikov 2013a, b).

Structure of the tourism system provides connections between elements. Links within the tourism system are based on social relations, which arise, change and cease regarding the several material and intangible profits of people consumed when moving outside their normal environment, as well as regarding the organization of tourist travels.

Allocation of ecological tourism as a separate species in tourism system is considered generally accepted. The role of ecotourism is assigned as an important component of sustainable development in the natural territories. Thus, ecological tourism must be considered in the classification system of modern tourism.

Ecotourism, as a subsystem of tourism social institution, is a social education in which social relations and interactions with the participation of subject carriers on social relations (ecotourists, travel companies, specially protected the natural areas) develop. Ecological tourism performs various social functions in society and meets individual social and environmental needs.

At the present stage of development in ecological tourism there are processes of characteristic formation, characterizing ecotourism as an independent subsystem of tourism:

- social relations and interactions are developing;
- subject-matter carriers of social relations—ecotourists, representatives of the tourism industry and specially protected the natural areas;
- the values of ecotourism in society are promoted;
- Russian ecotourism is gradually being integrated into the structure of international tourism organizations and associations.

But many important elements of ecotourism are still institutionalized:

- legal framework is not sufficiently developed in many countries;
- existing statistical record does not make it possible to assess the contribution of ecotourism to the socio-economic sphere of the region;
- insufficient training of tourism workers in the field of ecotourism;
- ecotourism development is not accompanied by an adequate system of environmental education and education of ecotourism participants (Zhukov 2016).

In conformity with the need for planning growth and territorial organization of tourism, tourism activities, and all tourism types in general as a spatial socio-economic phenomenon, it was essential to identify separate forms and types of tourism in several directions. Thus, the classification of tourism is necessary because it allows you to solve a number of problems concerning development and territorial organization of tourism and its infrastructure, to determine the demand for certain types of tourism services and, on this basis, to develop plans for the development of tourism's material and technical base.

Tourism can be classified according to various criteria: by purpose, nature, timing and duration of the trip, means of transportation, means of accommodation, the qualitative composition of trip participants, etc. But decisive importance still belongs to the purpose of travel. Truly, while travelling, tourists set themselves more than one goal. But depending on the individual needs of tourist, one of them becomes dominant (Yerdavletov 2000).

The most common classification of tourism is its division into types, categories, and forms (Table 1.4).

The classification is considered as general and is based on well-known types of tourism. But it does not give a complete picture of tourism, nor does it consider many factors. For example, ecological tourism is not considered in this classification.

Also, based on the same data, but supplemented with other types of tourism, the classification of other authors is given (Table 1.5).

Table 1.4 General classification of tourism [compiled by the author (authors) based on the materials (Durovich et al. 2003)]

Tourism types	Tourism categories	Organizational method	Travel duration	Intensity of tourist flows	Depending on the age category of tourists	Sources of financing	Purpose
1	2	3	4	5	6	7	8
Domestic	Within the country	Organized tourism	Short term	Permanent	Children	Commercial	Recreational
Inbound	National	Unorganized tourism	Long term	Seasonal	Youth	Social	Health
Outbound	International				Middle-aged		Cognitive
					Third-age people		Business
							Religious
							Ethnic
							Transit

Table 1.5 Classification of tourism [compiled by the author (authors) based on the materials (Kuskov et al. 2005)]

Feature	Description
1	2
Number of trip participants	Individual, group, family
Place of trip	National, international
Market	Domestic, foreign
Organizational method	Unorganized, organized, amateur (qualification)
Goals	Relaxation (rest), rehabilitation (sanatorium), recreational. Deployed: recreation and holiday, medical, business (business travel, congress, incentive tourism), educational, religious (pilgrimage and cognitive), ethnic, transit (according to the WTO, this includes other travel goals), shopping tourism, nostalgic and others
Age	Children, youth, middle age, third age, fourth age
Transportation methods	Automobile, bus, water (cruise), pedestrian, railway, aviation, bicycle, horse
Main activity	Business, educational, sports, environmental, congress, adventure, extreme, rural (agritourism), etc.
Mainly used the natural resources	Mountain, marine, forest, river, sun and beach, health
Territory	Continental, coastal, island
Financing sources	Commercial, social
Means of tourist accommodation	Hotel type, non-hotel type
Travel distance	Near, far
Travel duration	Short-term, medium-term, long-term
Intensity of tourist flows	Seasonal, year-round (permanent)

The classification is already more detailed, but still, it also does not fully disclose the tourism system.

Detailed classification of modern tourism is proposed by Yerdavletov (Fig. 1.1).

Depending on the purpose of tourist trip, excursion (cultural) tourism is distinguished in this classification, which in turn is divided into ecological tourism. And this again brings us back to the consideration of ecotourism within the cultural context and definition deduced to the author (authors) "Ecotourism is a type of tourism related to tourism needs, contributing to the preservation of environment and based on tourist demand developing within cultural context".

There are many classifications of tourism. They are different in compliance with understanding of this phenomenon, principles of construction, applied problems of classification, etc. It was considered that the Classification of Modern Tourism by Yerdavletov to be more detailed and applicable to us.

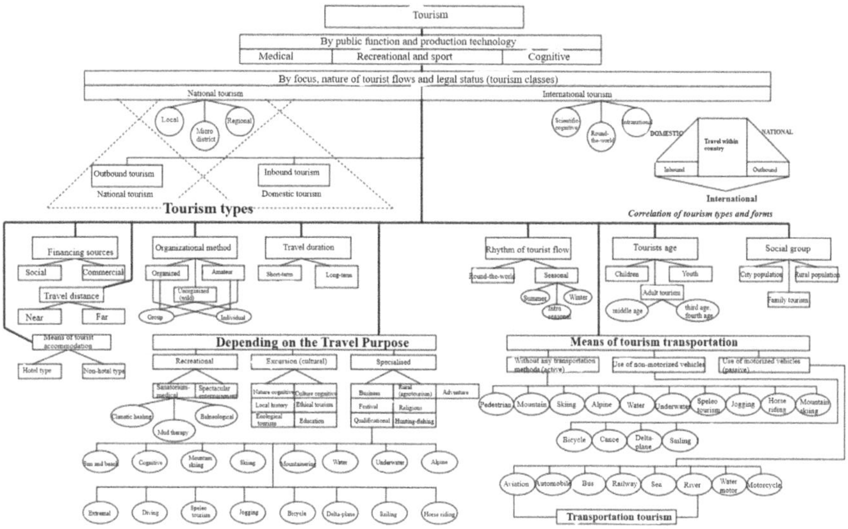

Fig. 1.1 Classification of modern tourism (Yerdavletov 2000)

In accordance with planning and organization of ecotourism, an vast range of activities can be attributed—from long scientific expeditions to short-term outdoor recreation on weekends. All this variety can be conditionally classified according to various criteria, in particular:

- by groups of travellers;
- by purposes and objects of travel;
- by type of tourism activity and others.

Furthermore, ecotourism has its characteristics that distinguish it from mass tourism, so ecotourists differ from mass tourists. Mass tourists, both organized and individual, often travel within known places, in most cases in their country and rarely move away from the so-called "beaten track". In contrast, ecotourists visit the natural areas that are little changed by humans, and they are more patient with limited amenities and primitive travel conditions.

Number of authors subdivide groups of travelling ecotourists according to their level of interest. For example Lindberg (1991) identifies 4 groups of ecotourists: members of educational institutions, scientific researchers; tourists visiting specially protected the natural areas; people, visiting the far corners of nature; those who believe that a trip to nature is an integral part of larger trip.

However, this classification cannot be considered absolute, as ecotourists may fall into other categories under different circumstances. Nevertheless, this classification is very useful, in particular, when planning ecotourism.

Similar typology of participants in nature tours is offered by Strasdas (2002). It divides 6 types of participants of nature tours: the natural tourism enthusiast; tourist interested in nature tourism; tourist occasionally interested in nature tourism; tourist interested in sports (adventures); tourist hunter (fisherman); tourist interested in nature and culture.

This classification already subdivides ecotourists, and has a focus more on nature-oriented tourism.

In accordance with type of tourism activity and specifics of occupations, scientific, cognitive and recreational tourism are conditionally distinguished (Table 1.6).

Allocation of all these ecotourism categories is relatively arbitrary. Sometimes it is very difficult to draw a clear line between different types of ecotourism, for example, cognitive and recreational. Also, other authors cited other types of ecotourism that complement this classification.

There are four types of ecotourism and ecotours: scientific, nature history tours, adventure tourism and travel to specially protected the natural areas (SPNA).

Table 1.6 Types of ecotourism according to the activity specifics of classes [compiled by the author (authors)]

Types	Meaning
1	2
Scientific	Scientific expeditions of scientists, summer field practices of students, and other trips involving the collection of scientific information about the visited region are sometimes referred to scientific ecotourism. Scientific tourism usually makes up a relatively small share in the overall ecotourism flow, but its role can be huge. In particular, it can significantly replenish information on poorly studied areas and objects. The information received may be useful not only for science development, but also for the further development of ecotourism on sustainable manner in the region. This type of tourism can contribute to the expansion of scientific and educational ties between different countries, lay the foundation for important international projects implemented through grants from charitable foundations
Cognitive	Cognitive ecotourism is "ecotourism in the classical sense". Its varieties can be considered birdwatching tours, whales or exotic butterflies, botanical excursions, archaeological, ethnographic and palaeontological trips, eco safari, tours for lovers of video and photography
Recreational	Main purpose of recreational ecotourism is outdoor recreation, although, of course, it also pursues certain cognitive goals. Such tourism can be active (being a kind of "adventure" tourism) and passive. Its active forms often include hiking, horseback riding and skiing, mountain climbing, caving, many types of mountain and water trips (rafting on catamarans and rafts, kayaks, canoes). Passive forms of recreational tourism can, for example, be considered: rest in camps, nearby hikes and walks, picnics. The category of passive recreational ecotourism can also include fishing and the collection of nature gifts, carried out on the principles of environmental sustainability

1. Scientific tourism. During the scientific ecotours, tourists take part in different kinds of nature studies, conduct field observations. For instance, ecotours are widely known associated with observing the behaviour of birds in Latin America, counting the number of whale populations in the Pacific Ocean. As a rule, tourist destinations in such tours are specially protected the natural areas (SPNA): nature reserves, nature reservoirs, national parks, the natural monuments. Scientific tourism expeditions include foreign scientific research expeditions, as well as field practices of students studying at the natural science departments of universities and institutes.
2. Tours on nature history. These are tours related to the acquaintance of surrounding nature and local culture. As a rule, such tours are a combination of educational, popular science and thematic excursions that run along specially equipped ecological paths. Most often they are also organized in the territories of reserves and national parks. This also includes tours of schoolchildren, during which the teacher and guide conduct excursions and speak about nature. This type of ecotourism is usually popular in Germany. Therefore it is also called the "German model for the ecotourism growth".
3. Adventure tourism. This view embraces all trips connected with active modes of movement and outside recreation, aimed at obtaining new sensations, impressions, raising the physical shape of tourist and achieving sports results. This includes such types of tourism as mountaineering, rock climbing, ice climbing, caving, mountain and hiking, water, ski and mountain ski tourism, canoe, horseback riding, mountain biking, diving, paragliding, etc. Many of these types of tourism have appeared recently and are considered extreme, as they are associated with great risk. However, this is the fastest development, profitable, albeit expensive, type of ecotourism. Adventure tourism is often called "heavy ecotourism" because the thirst of tourists for adventure here prevails over the motives of nature conservation. In the future, speaking of ecotourism, we will mean adventure tourism.

Adventure tourism is often identified with sport tourism (mountaineering, caving, paragliding, etc.) and active tourism, when tourists travel using the so-called active methods (on foot, by bike, boat, raft, etc.). Sport tourism as travel for playing sports or attending competitions includes types of tourism that are not associated with adventure or risk. For example, soccer teams going to training camps, or fans who follow their team for an away match. At the same time, adventure tourism includes types of travel without active means of transportation. For instance, jeep—tours over rough terrain on cross-country vehicles, or BASE—jumping with a special parachute from aircraft, steep cliffs, high man-made objects (arch bridges, TV towers, etc.) or deep-sea diving.

Present day, adventure tourism on the world market is not just adventure travel, but tours with elements of something unusual, exceptional for tourists.

4. Travelling to specially protected the natural areas. High attractiveness of unique and exotic the natural objects and phenomena located in the protected areas

attract many tourists. For instance, 48% of tourists arriving in Latin America have a purpose of travelling to nature reserves. Management of many national parks and reserves are defined as ecological excursions into a real show.

For instance, the Yellowstone National Park in the United States, where the duration of excursions is calculated in minutes and is linked with periods of geyser activity. Usually, the display of the natural objects, especially in caves, is accompanied by colour lighting, music, theatrical performances, showing scenes from the life of Aboriginal people. This type of ecotourism is most developed in Australia; therefore, it is identified with the "Australian model of ecotourism development" (Khrabovchenko 2004).

Types of ecotourism listed above do not fully cover the entire spectrum of ecotourists, but give a more in-depth picture.

Ecological tours can be classified according to many characteristics, but the two most important characteristics should be considered as the most significant—its purpose and object (Table 1.7).

Of course, the ecotour purposes and objects are interconnected; both main specific characteristics cannot be considered as absolutely independent classification basis.

If we speak about the classification of tourism according to the visiting objects criteria, the main objects of ecological tourism in its classical and environmental interpretation are considered as closely untouched the natural complexes or their united elements. Objects of cognitive or scientific ecotourism are separate notable biological types, the so-called types—flagships. A striking example of this kind is the mountain gorillas in Rwanda, for protection of which the Des Volcans National Park was created. Ecotourists are also attracted by unique objects of inanimate nature, geomorphological, hydrological and other features (individual mountains and canyons, caves, waterfalls, lakes and rivers, etc.), as well as palaeontological finds. Objects of ecotourism can be exotic plant communities and bioceonoses as a whole, for instance, tropical forests, flowering tundra in summer or desert in spring. In addition, ecotourism objects can be cultural, ethnographic, archaeological and historical sites, as well as the natural and anthropogenic (cultural) landscapes in whole.

Table 1.7 Classification of ecotourism by specifical features [compiled by the author (authors)]

Main purpose	Main object
1	2
Observation and study of wildlife with training in environmental knowledge	Botanical, zoological, geological
Emotional-aesthetic	Ecological-ethnographic or archaeological, ecological-cultural
Treatment with the natural factors	Agritours
Sports and adventure purposes	Speleological, water, mountain, etc.

Basic goals of ecotourism are to preserve and utilize the natural and cultural resources sustainably and to enable economic growth of local people. However, achieving the aims in ecotourism depends on whether they are environmentally and ecologically sustainable and economically applicable. To achieve these, a participative tourism planning is required (Kiper 2013).

Based on the above numerous classifications, the authors have compiled a general classification of ecological tourism, in which the role will be understood as such role in the tourism system (Fig. 1.2).

Dimensions of ecotourism in context of Sustainable development refer to the environmental, economic, and social aspects of tourism development, and a suitable balance between these dimensions must be established to maintain its long-term sustainability (Bhuiyan et al. 2012).

The most important contribution to nature protection is ecotourism development in Kazakhstan comes through the support of specially protected the natural areas, which play a major role in maintaining the ecological balance. However, the natural territories are at risk of pollution not only as a result of industrial development, but also because of an extraordinary influx of visitors, as a result of the intensive development of tourism in general.

Regarding the development of ecotourism in our country, it should be noted that, for example, in August 2019 the Ministry of Culture and Sports launched a pilot project on ecotourism development in three national parks of Kazakhstan. This is Sharyn, Ile-Alatau, Kolsai kolderi. Thanks to the introduction of the new model it is planned to increase the tourism flow from 1.3 million tourists per year to

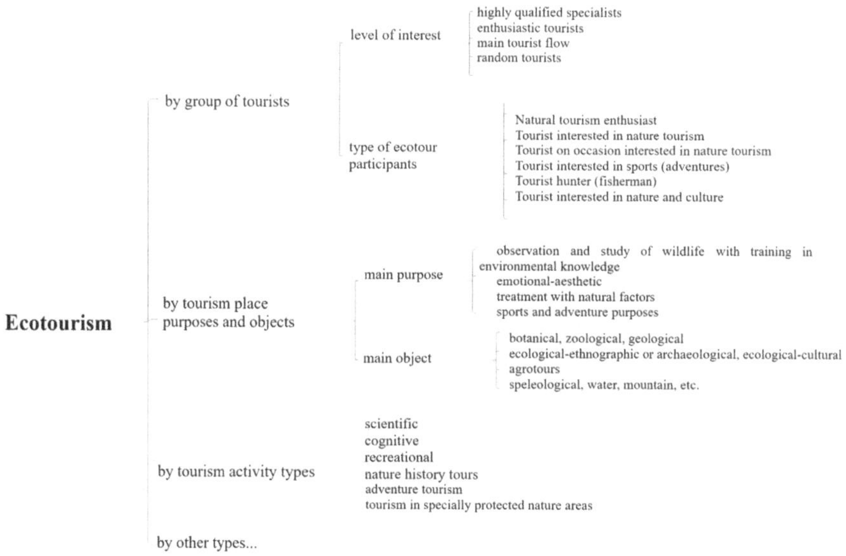

Fig. 1.2 General classification of ecological tourism [compiled by the author (authors)]

3.4 million tourists until 2025. These measures will create about 20 thousand new jobs and attract about 100 billion tenges of investments in the next 5–7 years. At present, there are 13 national parks in Kazakhstan, where there is a special regime ensuring the preservation of the natural resources, complexes and objects of the state the natural reserve fund. According to statistics, the annual growth of tourists in the National parks is on average 8% per year and according to the results of 2018 the number of tourists who visited these parks amounted to 1.3 million people. At the same time, ecological tourism development in the National parks does not take place at a sufficient level, due to the lack of necessary infrastructure that meets generally accepted world standards.

The so-called environmental crisis, causing irreversible processes, contributes to the rapid disappearance of the natural beauty (Boreyko 1996). Apparently, nevertheless, a time is coming when the beauty of wildlife will be protected only in nature reserves, national parks, areas of protected landscapes, landscape reserves, regional landscape parks and other protected the natural areas. If earlier the task was to protect areas where there were only unique landscapes, rare and endangered species of plants and animals, currently the task is much broader—to preserve and, if possible, restore and develop the aesthetic, recreational, cultural and historical qualities of the natural landscapes, which are the basis for ecotourism development.

In addition, it is necessary to reveal and show people all the uniqueness and diversity of pristine nature, which has been preserved in specially protected areas, and on its example, to cultivate in a person more careful attitude to it.

Role of ecological tourism in preserving biological diversity is also manifested in the fact that ecotourists themselves act as defenders of the territories they visited. As the experience of foreign countries shows, ecotourists are often ready to sacrifice time and make every effort to "advocate" for the nature conservation activities carried out in the territory they visited.

Many tourism organizations maintain in an environmentally balanced state the territories visited by tourists through the press, radio, television, organizing visits by scientists, ecologists, as well as seeking financial resources. As a result, ecotourists support the preservation of the natural environment, convincing their friends and family members to make such kind of tourist trips. Such support for our country would be more essential if it came from domestic ecotourists.

Concept of sustainable development in the Republic of Kazakhstan, strategic plans in the field of environment, developed by the Ministry of Environment and Water Resources of the Republic of Kazakhstan, indicate the future development of environmentally friendly types of tourism, such as ecological tourism.

In Kazakhstan, ecotourism development is not proceeding at such a huge rate, but the movement towards improving conditions for tourists has already begun. It is planned to develop a platform that will provide tourists with all necessary information about Kazakhstan and historical places of the country. The number of foreign tourists is planned to raise to nine million people.

For ecotourism development in our country, the Kazakhstan Tourism Association established the Information Resource Centre for Ecotourism (IRCE) in 2005. IRCE is a key point around which a programme to promote ecotourism based on the development of local communities in Kazakhstan is being implemented.

Main activities of IRCE:

- popularization of ecotourism ideas;
- promotion of community-based ecotourism into the tourist market;
- provision of information to tourists and visitors of the centre;
- booking tours and advertising support for project regions.

IRCE promotes ecotourism through the creation and development of new tourist destinations in different regions of Kazakhstan, their marketing and advertising (Table 1.8).

Ecotourism has been a petite segment in the tourism sector of Kazakhstan for many years. Only 10–12% of the total. On the international market Kazakhstan is interesting as a country for business offers and investments, but since the beginning of summer in 2016 the trend of more ecotourists' arrival has become visible.

Thus, trying to highlight ecotourism as the basis of national tourist product, Kazakhstan presented the idea of ecotourism in the international tourism exhibition China International Travel Mart 2018 in Shanghai. In 2019 finally began to solve

Table 1.8 Types of ecological tourism in Kazakhstan [compiled by the author (authors)]

Territory	Types
1	2
West Kazakhstan	Mountain Pedestrian Bicycle Excursion Historical Pilgrimage
East Kazakhstan	Ski Pedestrian Excursion
North Kazakhstan	Excursion Water Speleological Informative Historical
Central Kazakhstan	Mountain Water Excursion
South Kazakhstan	Mountain Water Excursion Informative Historical

the issue of roadside infrastructure, which is relevant for tourists. Since 2020, the compulsory registration of foreign tourists has been cancelled. Previously, foreigners who visited the country for rest and did not want to spend several hours travelling to register with the police risked deportation (Information Resource Centre of Ecotourism in Kazakhstan 2010).

Supporting the ecological tourism development in Kazakhstan, in 2020 UNESCO is launching a pilot project "Promotion of sustainable development in Kazakhstan through UNESCO facilities" in Almaty. The project will be implemented with the support of Kazakh Ministry of Foreign Affairs in five regions of the country: Almaty, East Kazakhstan, Zhambyl, Kyzylorda and Turkistan regions. The project aims to develop heritage-based tourism (cultural and ecotourism) at UNESCO facilities and their surrounding territory. It will focus on UNESCO World Heritage List sites, as well as sites with potential to be included in the UNESCO Global Geopark Network.

Ecological tourism for Kazakhstan is a new promising tourism sector, the main resources of which are aimed at protecting the natural ecosystem of different regions.

Government of the Republic of Kazakhstan in 2013 issued the concept of tourism industry betterment till 2020, which is in implementation. The concept envisages the development of 5 tourism clusters: Astana, Almaty, East, South and West Kazakhstan.

For ecotourism development on a cluster basis, the right approach is not enough. Sustainable tourism is not a kind of tourism, like beach, cultural and historical tourism. Such tourism philosophy pays close attention to every site, including sites such as beach, historical or any other tourism type. Therefore, a regional approach to its development is not applicable. Otherwise, such an approach would only be to preserve tourism facilities in the region. Thus, a sustainable approach should be defined in the Concept as a mandatory provision and general approach to tourism development in general.

Ecological tourism will develop sustainably when the domestic tourist follows the principles and rules of ecotourism, when all ecological trails and routes will be checked and classified by tourism categories, special routes and ecotours. These ecotourism routes should be protected from mass tourism and are based on the principles of sustainable tourism (Kairova et al. 2018).

Main objectives of ecological tourism development in Kazakhstan are as follows:

- to form and develop the natural and economic sector;
- reduce adverse effects on the ecosystem;
- develop the infrastructure of the region;
- restore destroyed the natural complexes;
- provide environmental education to the population;
- improve the socio-economic development of the region through tourism;
- rational use of resources in region;
- encourage the local population to protect the natural and cultural objects.

As a rule, the main objectives of global ecotourism are:

- environmental knowledge, improving the culture of relationships with nature, developing environmental standards of behaviour in the natural environment, fostering a sense of personal responsibility in each person for nature fate;
- restoration of spiritual and physical forces of man, ensuring good rest;
- environmental protection, minimizing damage to the environment;
- promoting the socio-economic development of individual regions and governments as a whole.

Ecotourism is an important source of income from nature conservation. The following sources of income are distinguished:

- payment for nature use in form of an admission fee on transport and accommodation use, hunting trophies, etc.;
- rent of individual people for the provision of services in form of food, housing, transport, guide;
- percent of sales to tourists visiting this territory;
- taxation of goods, services.

A number of authors proposes that ecological tourism can be considered as a potential means of preserving biological resources.

In any case, ecological tourism development should pursue the goal of preserving biodiversity, and at the same time bring a certain benefit. This will depend on a number of conditions, such as:

- territories visited by tourists should be characterized according to presence of interesting wildlife components that are accessible for sightseeing, that is, they must be unique or able to attract tourists;
- pricing policy should be developed that considers solvency of people, as well as collection of fees, the amount of which depends on prices of using ecotourism (Chizhova et al. 1989).

Ecotourism also plays an essential role in distribution of financial resources between countries with various income levels, in the differences on the natural environment of developing and developed countries. This averaging of lower and higher income countries is called the "equalization effect". A systematic approach to ecological tourism development should be aimed at solving the socio-economic issues of certain area.

Among all types of modern tourism, only ecological tourism can solve the above tasks. Being one of tourism areas, ecological tourism is the closest to environment. Therefore, it acts as the only connecting link in the formation of right relationships on system "man–nature" (Babkin 2008).

1.3 Pre-conditions for the Emergence of Ecotourism and Its Evolution

Tourism originated in the second half of the XIX century, as a peculiar form of recreation associated with displacement. It was based purely on economic relations —payment for organizing a trip for pleasure, knowledge and relaxation. The first tourist trips were carried out mainly in urban environments, man-made and cultural landscapes and set the task of introducing tourists to the historical and cultural attractions of a selected region. Over time, types of recreation appeared using the natural resources, primarily water. This includes swimming in the sea and inland waters, taking mineral baths, drinking mineral water, and mud treatment.

Over time, the leisure industry developed and became complicated by new types of tourism and recreation, which differ and are classified in several directions. So, there is a division of tourism by functional orientation, main occupation, degree of organization, method of movement, seasonality, composition of participants, duration and distance of travel, the amount of physical activity, etc.

Organized trips to get acquainted with the culture and nature of other peoples existed in the ancient world. So, in the VI century BC ancient Greeks and Romans travelled to Egypt, where they were attracted by ancient history, culture, unusual nature, magnificent Egyptian buildings. The first Greek tourist about whom reliable data are available is the "father of history" Herodotus, who described his many travels in nine books. His second book is dedicated to Ancient Egypt, which he visited after 459 AD. Even Aristotle, Democritus, Quintilian in their treatises on education recommended learning about nature in direct communication with it. Purposeful development of tourism as a means of physical development began to be carried out in the Renaissance (XV–XVI centuries), after centuries-old restrictions introduced by church into the sphere of spiritual and physical development of human being. Humanists sought to systematize education, considering various aspects of human life, identified the most effective means of forming human qualities. Thus, representatives of the Western European Renaissance X. Vives, M. Montel, T. More, E. Rotterdam spoke about the need to introduce an organized process of youth's physical education, giving a special place to the hiking use. The Vittorino de Feltre, teaching various tourist exercises in a school organized by him in 1425 in the city of Mantua, made many days of hiking with his students in the foothills of the Alps. Hiking trips as a means of healing people were practiced even in Jesuit seminaries.

During the Enlightenment (XVII century), Zh. Zh. Rousseau, G. Libli, and others saw in hiking trips a means on patriotic education of youth and strengthening their health, and teachings contributed to tourism development. Zh. Zh. Rousseau substantiates the importance of walking trips for solving health problems, developed theory of travel importance as an indispensable part of education, revealed the goals and objectives of travel and their relationship with other means of personality formation.

At the end of XVII century–beginning of XVIII century in European countries, some teachers used pedestrian walks and trips to nearby places of interest to work with students. They called such trips as excursions. In the second half of XVIII century in Western Europe, peculiar tourism centres began to be created.

Prerequisites for the subsequent growth of tourism were associated with the need to organize and conduct long trips to solve cognitive tasks of studying remote inaccessible corners of a globe. To improve motor abilities and skills of participants in such trips, their preliminary preparation is carried out, socio-psychological compliance and consistency are checked, the equipment necessary for travel is tested.

For many days, expeditions and excursions were applied in nature for a long time, and their goals and objectives were somewhat limited. Over time, the nature and content of trips and travels are enriched, and they are approved as effective means of physical education, which can solve educational and recreational tasks.

At the end of XVIII century–the first half of XIX century regional unions of travel lovers begin to form. The first amateur tourism organizations appeared in different countries at about the same time (in the second half of the XIX century) in the form of alpine clubs and societies. The first such club arose in England (1857), then in Austria (1862), Italy and Switzerland (1863), in Germany (1869), and by the end of the 70s. XIX century in France, Russia and other countries. In 1895, on the initiative of the Vienna newspaper "Arbeiterzeitung", the tourist association "Die Naturfreunde" ("Friends of nature") was founded. Soon, the first circles of society outside Austria-Hungary were organized, the spread of which caused the creation in Vienna in 1908 of the International Tourism Centre (Tourism: Study manual for students of pedagogical institutes on specialities 1988).

Ecological tourism has emerged relatively in recent years on the wake of "green" movement and has become widespread in the West due to the concept of countries sustainable development. Ecological tourism today is one of the fastest growing tourism types.

Along with the development of organized tourism, it was travel to the nature world that became one of its first directions. In the middle of the last century Thomas Cook began to organize veils into the Scottish Mountains, and in 1863 offered the first commercial ascent to Mont Blanc. There was a big interest in such travels facilitated by English romantic writers, who celebrated in their works the beauty of wildlife and delightful world of alpine peaks, valleys and meadows. In the America, travel to the nature world has also become popular thanks to literary descriptions carried out by well-known the naturalists and scientists who took part in scientific expeditions such as M. Lyusek, V. Clark, D. Odyubon.

Origin of the tourism knowledge system as a recognized scientific direction is attributed to the period of 1920–30 years (Hall 2005). It was at this time that tourism scientific research papers appeared.

Awareness of tourism was first systematized in the 1930s, and this was the research of historians. Then tourism knowledge was expanded by the studies of geographers, and then economists (Pemecky 2010). During 1920–30 years, i.e. 20–30 years before the official statistics of tourism appeared (in 1950), the geography of tourism received coverage and systematization. Researchers A. Carhart, A. Leopold, C. McMurray, D. Jones and others in 1920–30 years made a travel map and described the Anglo-American and European tourism traditions. In addition to geographical research, tourism knowledge has been forming since the 1920s. In social sciences, and also in the Anglo-American context.

Initially, there was neither a specific list of disciplines that systematically reveal the essence of tourism phenomenon, nor clear boundaries between academic disciplines based on the subject of tourism research. Borders were mobile, and disciplines interacted and echoed among themselves. And today, such boundaries are shifted when a new discipline appears that takes over a place within the pre-existing division of academic space (Johnston 1991).

Formation of clearer boundaries between research territories and academic subjects in the field of tourism has led to the rapid growth of tourism research and knowledge, as well as research and educational institutions of tourism.

In 1929, the Tourism Research Institute at the Higher Trade school was established in Berlin. The objectives of an institute were the study of tourism not only from the geographical and economic, but also from organizational, sociological, legal, meteorological and medical sides.

In Poland in 1936–1939 there was a Labouratory of tourism at the Institute of geography of the Jagiellonian university in Krakow and the Study group under the auspices of Tourism support league, headed by S. Leshitsky, who wrote about the complex of economic, geographical, statistical, legal, cultural and social problems related to tourism (1932).

One of the fundamental works of modern tourism science is considered the work of R. Glucksmann, "General information on tourism" (1935). He deduced the "law of opposites" (attractiveness for tourists to landscapes that differ from their places of permanent residence), exploring the recreational assessment of landscapes (Kuserova and Kovshov 2010).

Tourism Research Institute was opened in Switzerland, Bern in 1941, headed by Kurt Kranf, and Walter Huntsiker organized a seminar on tourist traffic at the St. Gallen Higher School, which in the 1960 was transformed into the Institute of tourism and transport economics. These scientists have made a great contribution to the comprehensive study of tourism.

In 1950, the German economic institute of tourism in Munich was founded in Germany, and in 1952, the Institute of tourism in Frankfurt am Main.

International organizations have played and continue to play a large role in expanding tourism research: IUOTO (International union of official tourism organizations, created in 1947, now UNWTO), IASET (International association of scientific experts on tourism, founded in 1951), International academy of tourism in Monte Carlo.

In Poland in 1956, the Institute of legal and economic problems of tourism was established in Warsaw. In France in the 1960s, the Institute of research and higher education in tourism of the Sorbonne University in Paris and the Centre for tourism studies of the Aix-en-Provence University were established.

However, the very first, oldest to date, tourist education schools appeared in Europe since the end of the XIXth century: Switzerland, Portugal, France, England, Scotland, Austria, Germany, and Holland.

Speaking about the prerequisites of ecotourism development and choosing the indicators characterizing them, it can be noted that, objects and goals of tours are far from identical in the main categories of ecotourists, to which one can refer:

1. Former amateur and sports tourists—participants of trips;
2. Lovers of green rural and summer life;
3. Fungi, some fishermen and hunters, etc.;
4. Guides of pupils, partly students;
5. Classical ecotourists;
6. Autotourists, friendly to the nature;
7. Middle-class members who value the environmental friendliness of recreation.

Several trends contributed to the ecotourism concept. Firstly, tourism has become one of the largest global economic activities, which has made clear the need for rational organization capable on providing real financial support for the protection of nature and increasing the essence of those the natural areas that should remain unchanged. Secondly, it has become clear that the success of environmental action is inconceivable through restrictive measures that affect the interests of local population. Finally, the change of priorities concerning city residents and their desire to rest in the corner of undisturbed nature contributed to the formation of demand for active-cognitive tours, which has "ecological" orientation.

Emergence and formation of ecotourism idea is accepted to refer in the late 1980s. By this time, many examples of the negative effects concerning unregulated tourism development had already accumulated. The most impressive and widely known are the nature degradation on a number of the largest national parks in the United States, recreational areas in the Alps, accumulation of rubbish mountains on Everest in Nepal and the Trail of the Incas in South America, etc. Modern scientists note changes in the behavioural environmental practices of the population in Europe and the United States during this period. Recreational degradation of the natural areas has led to the need to introduce a system of measures to protect nature into tourist practice. In this regard, it must be emphasized that the ecotourism idea has evolved inextricably from the concept of sustainable development.

During the period of 1970–1988 years, Western (including American) societies experienced a "cultural shift", which caused changes in the consciousness and value orientation of the population towards the dominance of the post material values over material ones. In the system of values the dominants were self-expression, self-improvement, also including the field of environment (Inglehart 1990).

In the early 1980s, the first scientific studies on negative tourism effects concerning the nature of recreational areas began to appear in the press. Criticism of "hard tourism" has increased. Home of these publications was Europe—densely populated, industrialized, leading in the development of tourism industry. In 1980, the West German magazine "Theo" published an article by the German futurist R.

Jungk "How many tourists per hectare on the beach?". In 1981, H. Weiss's book named "Peaceful destruction of the landscape and measures to save it in Switzerland" was published. Scientific development of this direction has led to the emergence of concepts such as "ecological tourism", "soft tourism", "green tourism" (Mozgovaya 2004).

The XX century brings fundamental changes in both focus and scale of travel to nature. At the beginning of the century, hunting safari to exotic countries was considered especially prestigious in order to get out the largest "trophy"—elephants, rhinos, etc. In the 1950 years, photo-safari is becoming more popular than hunting tours. Since the early XXth century, a number of causes have marked changes in the priority of recreation and tourism.

Firstly, development of production forms and living standards have contributed to the creation of spare time and have stimulated the growth of recreation and tourism industry. Secondly, increased urbanization and techno genesis have led to the separation of man from nature, so that preserved or intact the natural landscapes have acquired independent value as an object of tourism and meeting the recreational society needs (Kolbovskiy 2006).

After World War II, the tourism industry is spreading around the world. Technical revolution in transport and communications has made the most remote the natural corners accessible to mass tourism. In the seventies, the travel rich people is replaced by tourism of population with medium and even low wealth, youth tourism as hiking and horse hiking, mountain climbing, skiing, alloys on rivers is developing.

The first definition of ecotourism was given in 1983. The definition itself is closely related to terms used by the UN in the 1970s, such as "betterment without destruction" and "eco development", that is, environmentally acceptable development with the least negative impact on environment.

Term "ecological tourism" was proposed by Mexican economist-ecologist Hector CeballosLascureain in 1983. At present, in both professional and academic environments, there are several idealistic definitions of ecotourism similar in meaning, which have already been given by the authors earlier.

It is believed that the term "ecotourism" was firstly used by Miller in 1978 as one of the options for sustainable tourism development, but there are also claims that the concept of "ecological tourism" was firstly used as early as 1965 by researcher Hetzer. However, most researchers agree that the term was first mentioned by Mexican ecologist Hector Ceballos-Lascurain in one of his papers in 1983. In his original formulation, ecotourism is "travel to relatively intact corners of nature to familiarize and study nature, samples of wild flora and fauna, and cultural attractions located in these territories". In 1993, the original definition was revised by the author. The final formulation of researcher, ecotourism is "environmentally responsible travel through relatively undisturbed the natural areas with the aim of familiarizing and studying nature and related cultural attractions, contributing to the environment preservation, causing minimal damage to nature and creating socio-economic benefits for the local population by actively involving them in the tourist process". In 1996, this definition was approved by the International Union for Conservation of Nature (The International Ecotourism Society 2019).

Modern historiography does not offer a single interpretation of the term "ecotourism". On the contrary, diversity of approaches to the definition of phenomenon presupposes even greater reorientation of the issue, which in turn leads to the emergence of various "superstructures" combining the concepts of ecological, adventure and cultural tourism. Thus, the emphasis of modern research focuses mainly on the notion of ecotourism itself, as well as its social and economic impact on regional economy.

The first steps towards realization of sustainable development idea were taken by the creation in April 1968 of the Roman Club, which brought together more than 30 leading and authoritative scientists and sociologists. In the future, the concept of sustainable development was discussed and supplemented in the Rome Club (1972) reports, at the First United Nations Conference on environment and development (Stockholm, 1972). However, it is believed that the concept of sustainable development was first mentioned in 1980 in a study by the International Union for nature conservation. Then in 1987 the UN report named "Our common future", prepared at the initiative of the Commission on environment and development (Brundtland Commission), was published, which called sustainable development as "the basis for global strategy aimed at bringing to common denominator of economic, environmental and social objectives". There was a definition of sustainable development in report of the Brundtland Commission: "development at which the needs of present are met, without putting at risk ability of future generations to satisfy the requirements". At the same time, sustainability in tourism implies a positive overall balance of ecological, socio-cultural and economic impacts of tourism, as well as the positive impact of visitors on each other.

In the early 1980s, the first scientific studies on negative effects of excess tourist flows into recreational areas began to appear in the press. Not surprisingly, the homeland of these publications was Europe—a densely populated, industrialized, tourism-leading part of the planet. In 1980, the West German magazine "TEO" published an article by R. Junk, "How many tourists per hectare of beach? The word in defence of "soft" tourism". In 1981, H. Weiss's book named "Peaceful destruction of the landscape and measures to save it in Switzerland" was published. Development of this direction led to the emergence of concepts such as "soft tourism", "green tourism", which included recommendations on the conscious, environmentally competent use of the natural areas for recreational purposes. The mid-1980s was characterized by the intensification of activities to create new protected the natural areas, the development of recommendations for organization of tourism activities in these areas, and legislative provision of environmental protection.

In 1990, ecological tourism was granted official status, and it acquired the right to convene its own annual international symposium, "Annual World Congress on Adventure Travel & Ecotourism", and to establish its non-profit organizations.

In Western societies, the consumer lifestyle has been replaced by a sustainable way of life, the purpose of which was to create conditions for harmonious social, the natural and economic development of society (Dunlap 1993). The essence of this way of life is reflected in the concept of "sustainable development" adopted at the UN Conference on environment and development in Rio de Janeiro (1992). In this interpretation, ecological tourism is not the objective, but a process of balancing human needs at different levels of social organization. Thus, the post material installations and greening of the lifeway took a strong position in social consciousness after the economic recession in 1980.

Almost simultaneously in the Old World and in the New World, the idea of tourism compatible with environmental and social requirements was formulated, that is, responsible to nature, promoting its protection, enhancing the ecological culture of travellers, performing an educational function that is carefully related to traditional cultures and society of the local population.

In the late twentieth century, the correct consumption problem of the civilization benefits reached its peak. Much attention was paid to this topic not only by scientists (for example, J. Bodriyar "Society of consumption", 1970), but also by politicians. Thus, the US Vice President Albert Gore, while still a senator, wrote in his book named "The Globe in equilibrium" (1992): "Number of material goods is now higher than ever before, but the same number of people feeling the emptiness of their lives".

Because of change concerning priorities in tourists aspiration many special branches of tourism disciplines (tourism economy, recreational geography, tourism ecology, the biosphere and tourism, pedagogics and psychology of tourism, tourism management, safety of tourism, information technologies in tourism, recreation, tourism phenomenology, etc.) in a varying degree note the growing appeal of ecological tourism.

The number of publications, conferences and meetings devoted to the rational use of the natural, cultural and historical tourism resources has also increased as the problems have become relevant. Only in 1999, more than 400 papers on the subject were published abroad. An intense search resulted in the concept of sustainable tourism in the twenty-first century, developed in 1996 by UNWTO, the World Travel and Tourism Council and the Earth Council named "Agenda 21 for travel and tourism industry". The agenda was addressed to national tourism administrations, tourism organizations, as well as consumers of tourism services. The concept is based on "Agenda 21", which is a comprehensive program of action adopted by 182 governments at the United Nations Conference on environment and development dates from 14 June 1992.

According to "Agenda 21", sustainable development refers to a process that takes place without affecting tourism resources. This is achieved by resource management resuming at the same rate as is used, either by switching from slowly renewable resources to rapidly ones. Sustainable development is based on a combination of three pillars: sustainable social, economic and environmental development.

Development and implementation of Local Agendas—21, i.e. regional or local sustainable development strategies, was proclaimed as one of the ways to achieve sustainable development in the conference outcomes. In 2002, during preparation for the Johannesburg World Summit on sustainable development, the scale of such work worldwide was assessed, and it was found that some 6500 such initiatives had been developed and implemented (or were under way) in 113 countries in the past decade.

An important step in the evolution of ecological tourism was the proclamation by the United Nations of 2002 as the International Year of Ecotourism, with 132 countries participating in the World Ecotourism Summit held in Quebec, Canada, from 19 to 22 May 2002, under the auspices of the United Nations Environment Program (UNEP) and the World Tourism Organization. Along with it the Quebec Declaration on Ecotourism was adopted, which contains recommendations for ecotourism development in various countries. The summit recognized that "ecotourism is based on sustainable tourism, considering its impacts on the economic, social and the natural environment". In addition, the Quebec Declaration specified the following highlighted principles that distinguish ecotourism from the broader concept of sustainable tourism:

- Active promotion on the preservation of the natural and cultural heritage;
- Involving local and indigenous communities in the planning, development and implementation of ecotourism activities that contribute to their well-being;
- Explaining to tourists the significance of the natural and cultural heritage of the visited tourism centres;
- Focus on individual travellers and organized travel groups of small size.

A significant development for the further betterment of ecotourism was the World Ecotourism Conference, held in Oslo, Norway, from 14 to 16 May, which marked the fifth anniversary of the Quebec Summit. The conference was organized by the Tropical Institute of Ecological Sciences (TIES), the Norwegian Ecotourism Society and the United Nations Environment Program (UNEP). There were 450 delegates from more than 70 countries attended. As a result, the achievements and challenges of ecotourism since 2002, the joint work of national and regional ecotourism associations and professionals around the world were reviewed, and specific techniques and practices were identified to enhance ecotourism's contribution to the conservation and sustainable development of territories by:

- Recognition of the ecotourism value in local sustainable development;
- International agencies, governments and all ecotourism-related organizations;
- Maximize the potential of developed ecotourism enterprises as the main economic force on preservation of all types concerning the natural and cultural heritage;
- Supporting the viability of ecotourism enterprises and activities through effective marketing, training and study;
- Highlighting the most pressing challenges facing ecotourism to strengthen its sustainability.

Thus, the ecotourism concept, which originated in the 1970s–1980s, has passed a number of development stages—from the "elite" and little familiar for most people type of rest to the internationally recognized and actively developing course of the tourism industry. Many non-profit organizations in the field of ecotourism have emerged, and specialized thematic symposia, conferences, congresses and exhibitions have been held on a permanent basis. Major development in ecotourism was the proclamation of 2002 as the International Year of Ecotourism and its World Ecotourism Summit, followed by the World Ecotourism Conference in Oslo 5 years later. The world ecotourism community has grown significantly and continues to grow, which gives reason to speak of serious steps in strengthening the position of ecotourism in the world travel industry market.

The World Tourism Day, which has been held at the initiative of UNWTO since 1980 every year on the 27th of September, serves as an important event to promote and develop the social, cultural, political and economic values of tourism (since the UNWTO Charter was adopted on that day in 1970, and the World Tourism Conference in Manila, which adopted the Manila Declaration on World Tourism, was inaugurated on that day in 1980). Every year theme of Day changes, for example, in 2002—the International year of ecotourism, in 2008—"Tourism and climate change", in 2013—"Tourism and water: protecting our common future". Themes of World Tourism Day reflect humanity's global goals and tourism contribution to their achievement.

Increase on ecotourism popularity in the world is due not only to environment deterioration, but also to acceleration of changes in human environment, the increase in nervous and psychological loads. For example, the World Tourism Organization declared the 2020 year as for rural and ecological tourism. All its events in 2020 UNWTO will be held not in capitals, but in provincial cities and regions.

Relatively in recent years, environmental activities have come to be defined as actions that are carried out to change (usually to improve) the environment. Today, sometimes the main motive for eco-trips is the question of tourist's identification and motivation.

Thus, ecological tourism, which was born at the beginning of the XXth century as an "elite" type of tourism (designed for a wealthy customer), has grown to an actively developing direction of the natural tourism. Non-profit organizations in the field of ecotourism have appeared, and specialized events are held on a permanent basis. A generally recognized feature of ecotourism organization is to ensure minimum impact on the natural environment, to create a network of ecological hotels, to provide tourists with environmentally friendly nutrition. Assistance in ecological tourism development has brought some innovations to modern tourism, primarily the reduction of costs and excess consumption.

To better understand the history of ecological tourism development, the authors have drawn up a chronology of ecotourism development because of above data (Table 1.9).

According to statistics, 1.3 billion international tourist arrivals were recorded in the world in 2017, which provided about 10% of world GDP. Moreover, according to some data, today the share of ecological tourism covers up to 25% of the world tourism market (WTO 2001).

Table 1.9 Chronology of ecotourism development [compiled by the author (authors)]

Time	Events
1	2
2nd half of the XIXth century	The birth of tourism
1863	Travels organized by Thomas Cook
1880	Formation of ecotourism idea
XXth century	Change in priorities. Most definitions of ecotourism have been given
1920–1930	Emergence of tourism knowledge system
1980 years of XXth century	Emergence and formation of ecotourism idea
1970	UNWTO Charter
1980	World Tourism Day
1980	World Tourism Conference in Manila
1970–1988	"Cultural shift" towards ecological tourism
1983	The first definition of ecotourism
1980	The concept of "sustainable development" was given
1990	Official status of ecotourism
The end of XXth century	Shifting priorities towards rational resource management
1996	Concept of sustainable tourism in the XXI century
2002	International Year of Ecotourism
2002	World Ecotourism Summit
2007	World Ecotourism Conference in Oslo
2020	Rural and Ecological Tourism Year

Development trends in 2019 were such that ecology and its protection became for many ones of the main factors in holiday planning. By choosing less environmentally harmful modes of transport, the younger generation plans to travel on bicycles, electric cars, and electric trains. They change their route and give up planes, trains, cars that leave a huge fraction of carbon emissions. Hiking tourism is rapidly gaining momentum as a symbol of 2020.

Virtual reality technologies, the desire of more people to know their ancestors, and the struggle for ecology will have a big impact on tourism in 2020.

1.4 Factors for the Development of Ecological Tourism

Modern economic conditions make it possible to position tourism as the most important branch of national and world economy, alongside high level of income and intensive rate of development. Among the reasons for high profitability and growth rate of industry are the significant demand for travelling into different directions and efficiency of tourism products and services. In the economy of most countries, tourism industry allows stimulating not only economic, however also social development of regions, and moreover plays a significant role in increasing state budget revenues.

One of the promising and developing destinations of tourism industry is ecological tourism.

Ecological tourism grows by 25–30% per year (WTO 2012) and has significant potential for further development. According to the WTO "Tourism Vision 2020" forecast, the number of international tourist visits will increase to 1.6 billion units (WTO 2012). Pursue the principles of sustainable development on global concept; discussions about the possibility or impossibility of achieving a fuller development of tourism continue to be unabated.

For example, a number of authors have made ecotourism crucial in combating poverty and hunger in low-income countries, particularly in Africa (Santarem et al. 2018). It is believed that there are not many successful examples of ecotourism in the world. In this chapter we will consider the main factors of ecological tourism development.

Ecotourism has many positive factors. These include the following:

1. Ecotourism aims to maintain environmental sustainability.
2. Ecotourism arranges financing for the protection of the region environment and support of residents, furthermore contributes to regions betterment where it is popular, both economically and socially. Tourism aims to raise people's awareness in the field of ecology and environmental education.
3. Ecotourism today acts as an alternative to other uses of surrounding the natural objects that harm nature. The negative effects of ecotourism, which are socio-cultural and ecological, are minimized (Spasibenko 2015).

For ecotourism, familiar factors such as the number of travellers and motivation are critical, but for a given period of time not all of them are decisive and capable for ecotourism. The result of ecological tourism should be not only the environment study, but also, above all, the change of attitude towards nature from consuming to careful.

There are five criteria for environmental tourism that can be identified. Ecological tourism should be:

(1) Nature-oriented and based on the use of predominantly the natural resources;
(2) Non-damaging or minimizing damage to our habitat, i.e. ecologically sustainable;
(3) Aimed at environmental education and awareness, at forming relations of equal partnership with nature;
(4) Caring for the preservation of local sociocultural sphere;
(5) Cost-effective and sustainable development of the territories where it is implemented (Yerdavletov and Aliyeva 2013).

Crouch and Ritchie suggested that the competitiveness of the tourist spot was attributable to four main factors:

(1) support factors and resources including firm infrastructure, accessibility, promotional resources, and enterprise;
(2) core resource and its attraction including topography, history and culture, market relationships, integrated activities, special festivals and constructions;
(3) management of the tourist spot including resource management, marketing, organization, consultation and service;
(4) qualifying criteria including local and surrounding environment, safety, and cost (Crouch and Brent Ritchie 1999).

Mihalic explored the competitive advantage of a tourist destination from the perspective of environmental management and the findings showed that tourist spots would be affected by environmental impact as well as environmental quality, and that the competitiveness of tourist spots could be boosted by effective marketing activities (Mihalič 2000).

Development of ecological tourism, as well as tourism in general, is determined by two most important factors: the natural-recreational and historical-cultural. Rich cultural and the natural potential allows countries that are not even among the most economically developed to gain a strong position in the world tourism market. However, the active public policy on tourism development is a prerequisite for this (Kosolapov 2005).

Ecological tourism development is due to a number of environmental factors in the country area or its region, which are a combination of purely the natural and the natural-anthropogenic factors; moreover, these are not means of work, objects of consumption or sources of energy and raw materials. Such factors include the level of purity concerning consumed water; forest area and variety of green spaces; moistening mode; landslide hazard; type of soils; land relief; wind and temperature conditions; seismic of the territory; pollution of the air basin; noise, radiation and other anthropogenic pollution, etc.

Socio-economic essence of ecological tourism implies the economic benefits of protecting the natural and cultural property. Currently, however, this interpretation of ecological tourism is not always implemented practically. Modern large tour operators, as well as medium and small tour firms, organize the sale of ecological tourist products along with ordinary excursion, beach and other types of recreation, at the best listing there is a small part of benefit for the needs of specially protected the natural areas (SPNA). It is quite obvious that the issue solution should be sought at the state level, by creating an appropriate normative legal base in each region regulating the activities of tourism companies, as well as aiming at the SPNA growth.

Ecotourism has become the main use of the natural resources within the SPNA. It makes conservation economically profitable. Ecotourist visits cover a wide range of purposes, from conducting scientific research to short-term weekend holidays. Some ecotourists are attracted to unique plant communities, such as flowering tundra in summer or non-living nature objects (canyons, caves, etc.). Others willingly spend their spare time watching crocodile nurseries or sea turtles laying eggs on the beaches of the Mexican coast. Thirds—keep records of the number of birds and mammals in remote corners of planets. Observations of large predators and hookers in East Africa and Asia are of particular interest. Basis for the ecotourism development in every country is unique the natural resources and specially protected the natural areas (SPNA), biodiversity, the natural sites on the UNESCO World Heritage List.

Present day, biological diversity is preserved and restored at state level through the establishment of network on specially protected the natural areas (SPNA)—land plots, water surface, and subsoil with special environmental, scientific, cultural, aesthetic and recreational significance, where the natural complexes and their components are protected. At the same time, among protected the natural areas there are reserves distinguished by outstanding, unique, universal value in terms of science, protection or the natural beauty. Their preservation is essential not only at the national level, but also at the global level, as world the natural heritage sites.

The allocation of protected areas and selected the natural objects, including those significantly transformed by human beings, is important for the solution of many issues and, above all, is necessary for the preservation of biodiversity, recreational resources, educational and studying purposes. Along with increasing human impact on landscapes, the problem is becoming increasingly significant. Network formation on protected the natural areas is a continuous process involving the identification of unique the natural objects, the creation of the natural-man-made systems that perform environmental, recreational or other functions that are crucial to humans.

But, the informed attitude towards these territories not only by specialists, but also by the population of the region as a whole depends on awareness level of their importance, uniqueness and vulnerability of these facilities. Demand for this information is largely determined by the ecotourist potential of these territories. It depends on the following factors:

- location of specially protected areas;
- diversity (to a large extent biodiversity) and attraction of ecosystems and landscapes;
- bioclimatic and ecological friendliness;
- transport accessibility and distance from settlement centres serving as suppliers of customers, which determines the cost of travel;
- infrastructure, security and level of socio-economic development of the territory;
- fame and popularity of the region and route (Kosolapov 2009).

Indicators characterizing the features of these factors action, except for the latter, can be grouped into three blocks.

Possibility of ecotourism development is primarily determined by the natural characteristics of territory. The natural block includes indicators of comfort on the natural conditions, ecological situation and proxy indicator of landscape diversity, based on assessment of forestry, density of river network and roughness of relief.

Intensity, types and forms of ecotourism are closely related to the overall level of socio-economic development in region, which either hinders or contributes to the realization of the natural prerequisites for ecotourism development. Therefore, the second block (socio-economic) includes an integral indicator of investment attractiveness and population health quality.

The third indicator—potential tourist demand—is intended to reflect the location of the area relative to main settlements, which are the main suppliers of ecotourists.

The third block (ecological-tourist infrastructure) consists of indicators designed to assess the state of ecotourism infrastructure in the area. The level of its development characterizes the territory's ability to accommodate and serve tourists. In addition to such traditional characteristics of tourism infrastructure as density of accommodation facilities (specific for ecological tourism) and roads, the block forms the following indicators: availability of tourist personnel, density of museums (local history, the natural history, museums-reserves) and parameters of ecological-tourist infrastructure (Almukhamedova 2013).

Usually, any kind of tourism has a certain negative impact on the natural environment. In the case of ecological tourism, this impact is minimized by the presence of specially equipped routes on the natural park territory, which have been called "cognitive trails of nature", or "ecological trails" (Vorobyev and Ryashenka 1988).

Considering the interests of visitors, such trails may have different purposes, length, etc.

It should be noted that some ecotourists, after visiting protected areas, support the preservation of the natural diversity by providing certain financial resources. For example, in many countries of America, private nature conservation activities are practised in some countries of Africa.

Desire to preserve the natural habitat of wild animals leads to creation of numerous private nature reserves worldwide, as they contribute further to the conservation of the natural resources by the public, and moreover help to increase the size of some the natural species (Aigistova et al. 1996).

Zoning is the most effective tool for managing protected the natural areas. It is through zoning that tourism can be organized in a way that minimizes any disruption or invasion of wildlife. Acceptable and sustainable tourism will depend mainly on biological properties of the zone, vulnerability of the natural species, as well as economic and human resources necessary to organize protected area, the natural park and set of services provided to tourists. At the same time, specialists must assess the ability of visited territory to withstand the tourism loads. This makes it possible to consider the number of visitors, to compile statistics of visits, and if necessary to suspend the tourists flow in time (Niyazbayeva 2006).

Thus, the huge potential of ecological tourism as an economic and environmental activity can also be realized through public administration and regulation. As for ecotourism to bring financial and organizational benefits to wildlife conservation, there must be appropriate pricing policies that reflect the actual cost of services for tourists and cost of operating the natural sites. Experience shows that when the local population is involved in ecotourism projects, violations such as non-destructive tree cutting or poaching will be significantly reduced. At the same time, this engagement should not just be in the form of handouts and charitable aid, or even food supplies to schools, hospitals and social services funded from tourist revenue. It is important that the local population itself take an active part in eco-tourism activities.

Successful practices of interaction between the tourism enterprises and state include contributions from Australian tour operators selling tour products on a range of coral reefs and coral islands along Australia's northeast coast, the Great Barrier Reef. Organization of contributions is determined by the Australian government and its purpose is to protect this unique the natural complex.

Ecological tourism plays an equally important role in the redistribution of financial resources among countries with different income levels, in the quantitative and qualitative differences between the natural environment of developed and developing countries. Averaging of higher and lower income countries has been called the "equalization effect". Among the difficulties faced by ecological tourism, the main ones are chaotic organization of tourism activities, the lack of understanding on prospects of this type of tourism between most tour operators, the absence of unified marketing policy, weak state support, lack of knowledge about ecological tourism and its places of conduct among potential consumers.

Next task is to concentrate the efforts of tour agencies to develop ecological tourism in domestic and foreign markets, based on systematic approach to the problem of ecological tourism and its practical component.

A systematic approach to ecological tourism development is aimed at solving socio-economic problems of certain territory. Development of ordinary types concerning tourist recreation in certain territories affects the natural environment conditions in a not optimal way, which reduces the attractiveness of territory and tourist activity. Scientific literatures, studies on ecological tourism emphasize the complexity of processes on physical impact to the natural environment; the need to observe ecological tourism principles is aimed not only at preserving but also at increasing the national the natural wealth. Socio-economic problems can arise with the intensive development of conventional tourism types, which imprint not only on cultural peculiarities, established traditions, but also on formation of the legislative framework and language environment. These issues are particularly relevant in the context of world economy globalization, development of transnational companies, labour migration and other problems of modern society. It is quite obvious that the loss or even "erosion" of cultural and historical aspects reduces tourist activity, decreases the role of tourism business in country development or its individual territory.

It is necessary to highlight the peculiarities of transport and logistic processes in those regions where tourism development has a rather high level. Both international entry, exit and domestic tourism significantly increase the load on transport industry, including the raise in length and density of transport communications, number of operated airports, transport and transfer nodes, passenger traffic across all modes of transport, increasing the transport mobility of population, coefficient of aviation mobility among population. The need to solve these socio-economic problems shifts the focus in tourism activities to ecological tourism, which maintains a balance between the development of the natural territory and conservation, as well as reproduction of its resource availability.

Economic function of ecotourism is manifested in the fact that it directly or indirectly has strong influence on many other sectors of development concerning economy (transport, communications, trade, construction, agriculture, production of consumer goods, production of souvenir products, and development of hospitality industry).

From an economic point of view, ecological tourism can contribute to improving the infrastructure of settlements near environmental routes. At the same time, the local population can benefit significantly from ecotourism, improve its socio-economic situation by working as security guards for tourists or tourist groups, providing tourists with various consumer goods and food, and renting houses to tourists. It can be said that ecological tourism is one of the types of economic activities, as it can bring the state a large income by importing foreign tourists into the country a certain amount of foreign currency (Stanev et al. 1984).

In order for ecological tourism to have very positive impact on the country's economy and social sphere, as well as to be the priority of tourism industry, its concept must have three main aspects:

(1) tourist orientation towards consumption of environmental resources;
(2) preservation of the natural environment;
(3) maintaining the traditional way of life among population in peripheral regions.

In the end, ecological tourism will contribute to changing the consciousness and habits of people, nurturing in them a careful attitude to nature, thus contributing to preservation of the environment in the natural conditions. However, in order for ecological tourism to truly contribute the preservation of environment, it is necessary that local populations and local authorities clearly understand that benefits they derive from exploitation of territory's the natural resources as objects of ecological tourism must be closely linked to preservation of the natural territory. Meanwhile, this benefit must be appropriately distributed and used so that it is actually an incentive to preserve nature. Otherwise, if profits are misallocated to the so-called narrow circle, it will not encourage the local population to use resources of the protected area rationally.

Ecological tourism and related activities are thus entirely dependent on the natural environment conditions.

So, ecological tourism development determines the importance of systematic approach to the problems of using free time, promoting and creating conditions for widespread introduction of healthy lifestyles.

Based on examples of countries where ecotourism is significantly developed, ecotourism is also a factor:

(1) infrastructure using modern technologies in the sphere of energy consumption, water consumption and waste disposal;

(2) conservation of biological diversity;
(3) effective management of cultural heritage.

There is a need to understand the potential positive and negative socio-economic effects that may arise in ecotourism development and to take them into account during planning and management of tourism activities (promotion of economic development in countries and regions; stable currency inflows to country/region and others).

It is believed that these directions determine the vector of future ecotourism development, as it has long not meant ascetic recreation, many of tomorrow's ecological resorts will be a shelter of luxury, as ecotourism concepts are embedded in comfortable resorts around the world (Shmitz 2008).

Trends in environmental tourism are determined by the World Tourism Organization (UNWTO). According to the forecasts of this organization, ecological tourism is one of the five main strategic directions of development for the period up to 2020.

Ecotourism is gaining popularity worldwide. The UN World Tourism Organization (UNWTO) expects environmental, the natural, heritage-related, cultural and adventure tourism to grow rapidly in the next two decades, and global spending will grow faster than tourism as a whole. The International Ecotourism Society estimates that ecotourism will soon account for 25% of the global tourism market and generate $470 billion in revenues.

Tourism has become a global phenomenon in the world economy. Its mass nature poses a significant potential threat to the natural and socio-cultural complexes. At the same time, it became obvious that with rational organization, tourism can provide real financial support for the protection of nature and preservation of unique the natural complexes in their original form. It has also become clear that the success of environmental protection activities is unthinkable only through restrictive measures aimed, among other things, at the interests of local population. Tourism, on the other hand, can create conditions where care for the preservation of the natural territories and cultural heritage will become economically beneficial for the local population. Moreover, there has been a significant change in priorities and aspirations of tourists themselves. An increasing number of tourists in industrial-developed countries began to strive to visit territories with undisturbed nature. In contrast to the traditional beach-resort holiday, and there was an interest in adventure-cognitive tours. These are one of the main trends in ecotourism (Kudakayeva 2014).

However, it should be noted that the active ecotourism development has negative consequences. Compared to the past, in recent years ecotourists have seen a gradual decline in intellectual requests, social responsibility, environmental concern and political savvy. Elderly, rich and exhausted travellers prefer comfort to nature conservation. Modern tourists, though much more numerous, are generally much less interested in the subtleties of the archipelago's unique ecosystem and wish to hear only a brief historical and ecological account of the islands. These trends reflect the erosion of true ecotourism value—the movement from real ecotourism to "facilitated", "light-ecotourism" (Plemarom 1999).

According to expert estimates, the number of people concerned about nature is increasing. Environmental disadvantage creates a need for some people to change their attitude towards nature, for others to actively participate in its protection and preservation, for others—the most conscious—in recovery.

To understand the trends of ecotourism development in 2020, an international study of ecotourists was conducted by the modern retargeting agent Criteo. According to the results of this study, it is clear that Germany is a leader in ecotourism. The 22% of German travellers who participated in the survey came under the definition of "ecotourist": travellers who have given up travelling at least once in the past six months to reduce their carbon footprint. France, Spain and the UK as the countries with the highest percentage of ecotourists were also among the top five ecotourists (Fig. 1.3).

Taking into account the huge variety of overseas and home destinations for travel in Europe, furthermore the availability and comfort of train travel, one should expect a further increase in the proportion of European travellers choosing ecotourism.

Our study found that Generation Z and millennials were the most conscious travellers willing to change their habits to protect the environment. Out of all the travellers surveyed by Criteo, 62% of those who refused to travel in the last 6 months were Gen Z representatives or millennials (Fig. 1.4).

Ecotourists are markedly younger than the average age of travellers.

According to the Booking.com study, 54% of Gen Z representatives note that the impact of travel on environment is one of the decisive factors in choice of direction, and 56% prefer to stay in eco-friendly hotels/flats. Also, 87% of tourists choose

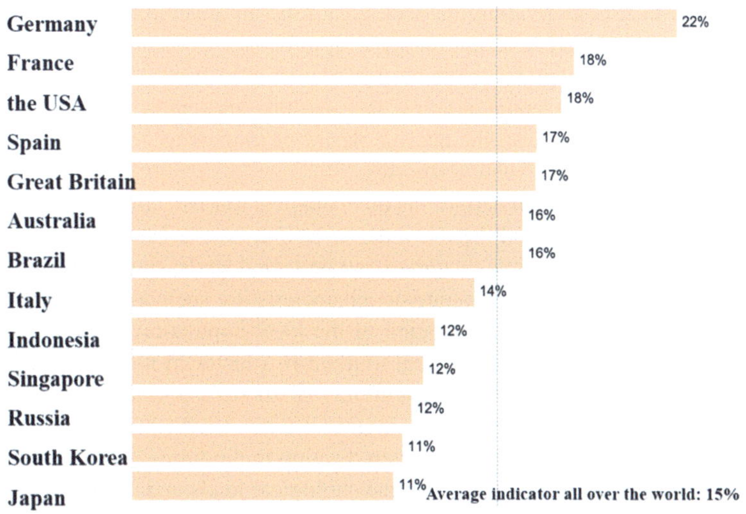

Fig. 1.3 Percentage of ecotourists by countries (Green Travel Trends 2020: Gen Z and Beyond 2019)

	Ecotourists	All travellers
Gen Z and millennials	62%	41%
Gen X and older generations	38%	59%

Fig. 1.4 Percentage of ecotourists by generations (Green Travel Trends 2020: Gen Z and Beyond 2019)

greener options if possible, and 67% are ready to overpay from 5% or more provided that their journey has minimal impact on the environment.

So far, ecotourists are still in the minority, but they are already of considerable importance. The Mandala Research study revealed that travellers choosing eco-friendly housing are delayed on average three days longer than other guests. According to a study by US travel website TripAdvisor, 30% of the world's tourists are ecotourists every year. In addition, 70% of tourists say that next year they are going to choose a safer type of tourism.

The main flow of ecotourists in the world is made up of the inhabitants on industrialized countries interested in nature and life preserved concerning the virgin corners of planets. Particular interest is an exotic nature and culture of Africa, Asia and Central America. Targeted policies of the host countries contribute to formation of a large ecotourists flow to Kenya, Tanzania, South Africa, China, Thailand, Ecuador, Costa Rica, as well as to New Zealand, Australia, etc.

Numerous factors motivating the ecological tourism development are the desire of people to get involved in nature during rest, to be able to breathe clean air, drink and swim in clean water, to consume environmentally friendly products, to get aesthetic pleasure from contemplating the natural landscapes. Reasons for the choice of ecotourist trips also include the desire of tourists to get acquainted with animals and plant world of new territories.

A prominent feature of all definitions on ecological tourism are given in previous chapters, and viewpoints, and there is a desire to link ecotourism predominantly to the natural, undisturbed or little disturbed and protected landscapes. However, some of these definitions indicate ecotourism characteristics such as concern for the social, cultural and economic well-being of the local population.

Thus, the very essence of ecological tourism is a factor in its development.

It should be noted that the current trends of modern society may affect the ecological tourism development in the future (Worner 1992).

First, public production will continue to develop in the future, which will lead to an increase in intellectual, nervous and psychological loads and, accordingly, raise the need to spend free time in more the natural conditions.

Secondly, the continued involvement of the natural resources in production will lead to further degradation of the natural environment. In this regard, the importance of ecotourism as a tool for removing part of the natural ecosystems from industrial and agricultural use will be strengthened.

Thirdly, the life of modern human beings is burdened with scientific and technical achievements. Therefore, it is necessary to achieve "simplification" of lifestyle, change of social values, including the spheres of recreation and tourism.

These processes are common to all countries of the world on varying degrees. However, to reduce their impact on ecological tourism development concerning the SPNA, a number of measures are needed at the local and state levels to improve ecotourism activities.

1.5 Models of Organization on Ecological Tourism Development in the World Practice

Global tourism industry has always been the largest and rapidly developing. It not only creates works and profit, but also contributes to the preservation of cultural heritage and ecological tourism.

When Hector Ceballos-Lascurain introduced the term "ecotourism" in 1983, there were more than thirty varying degrees related and interconnected definitions and terms. Sources of literature give a wide range on ecological tourism definitions: "green tourism", "non-technical tourism", "nature-oriented tourism", "adopted tourism", "adventure tourism", "agritourism", etc. Most significantly are the studies of German scientists, who divide tourism into two types—rigid and soft or technicized and non-technicized (Drozdov 2005).

For example, nature-oriented tourism is any kind of tourism that depends on the use of the natural resources in their relatively unchanged state, including landscapes, relief, waters, vegetation and wild animals. Unlike ecological tourism, the concept of "nature-oriented tourism" is based only on motivation of tourists (rest in the wild, familiarity with it) and the nature of their activities (alloys, trekking, etc.) and does not consider the environmental, cultural and economic impact of such travel. Therefore, the use of the natural resources in this type of tourism is not always reasonable and sustainable (it is enough to mention its types, such as hunting, motorboat travel, etc.).

Wildlife tourism and wilderness travel is sometimes identified as a form of nature-oriented tourism, which can target any wildlife species, from individual ones to communities and biocoenoses.

Ecotourism is often associated with adventure tourism. However, ecological tourism does not always imply an adventure component. On the other hand, not all adventure tours meet environmental criteria.

Green rural tourism, or agritourism, especially popular in the United States and Western Europe, is a holiday in rural areas (in villages, on huts, in convenient houses).

Green tourism implies the application of environmentally friendly methods and technologies in tourism industry.

In German-speaking countries, the adjective "ecological" is rarely used and definitions of "green" tourism industries are hardly seen. There, the terms as "Sanfter Tourismus", or "ecologically and socially responsible tourism" was the most widely used.

According to a number of authors, the concept of "soft tourism" is close to "green tourism" and characterized by environmental management, non-degrading resources, as some benefits from their development are used to promote and protect the latter. It covers a wide variety of tourism areas, from hotels, airlines and other modes of transport to tours of protected the natural areas. In this case, they are united by responsible attitude to environmental protection, in particular, the use of environmentally friendly technologies.

According to J. Krippendorf's definition, the main purpose of the new tourism model, originally described by this author as quiet, or calm tourism, is to provide long-term physical and spiritual recreation to as many people as possible in direct contact with the landscape, as well as to meet the landscape and needs of holiday forms of settlements, considering the long-term interests of residents.

Definition used by P. Haslacher identifies four "strategic" components of the ecotourism model, presence and interaction of which characterizes soft tourism: non-technological tourism, independent regional development, social and cultural responsibility. All these components are considered equivalent by the author. Thus, in Haslacher's definition, the priority criterion for soft tourism is its non-technical nature. In addition, according to Haslacher, soft tourism cannot develop in all tourist regions.

A similar-structure but different-content definition was proposed by K. H. Rohlitz. According to this author, soft tourism is a model, also consisting of four interacting elements: optimal rest of guests, "intact" landscape, that is, such guests come into contact directly, intact local sociocultural sphere and profit from tourist activity. The definition itself is formulated as follows: "Soft tourism is quantifiable moderate tourism, which, through remote integration, creates economic benefits for residents and ensures mutual understanding between guests and residents, and moreover does not cause damage to either the landscape or local sociocultural environment" (Hammer et al. 2007).

In international ecological tourism, the main areas generating demand are Western Europe and North America. By contrast, the territories most visited by ecotourists are in developing countries of Latin America, Asia and Africa, as well as Australia and Oceania. This distinguishes ecotourism from mass tourism, the vast majority of which flows from developed world to developing world. On the scale of the world tours geography, the German consumer with interest in nature-oriented and ecological tourism favours Costa Rica, which leads by clear margin from competitors. One of the reasons for this is the stable political situation in the country for many years.

Various models of ecological tourism organization have developed in the regions of the world. Until the middle of the last century there was no such thing as "mass tourism", to enjoy the natural beauty, to penetrate exotic corners of the Earth could be only outstanding "singles", obsessed with passion for adventure, having considerable energy and willpower. And basically, they were people very wealthy. Along with development of organized tourism, it was travel to the world of nature, i.e. nature-oriented tourism, which became one of its first directions. So, the "Australian" model of ecological tourism was born and began to develop.

Two models of ecological tourism vary on different requirements for their provision: Western European and Australian. Representatives of the first model include developed countries—Italy, Spain, France, Germany, etc. Requirements in this case to ensure ecotourism are very high, and it is connected with creating maximum comfort for human life in nature. For such a model it is necessary to build comfortable hotels, campsites, bridges, architectural structures, deforestation for supermarkets, parking and recreation houses, development of transport network, etc. Such tourism is expensive, but it generates high revenues. On the other hand, it leads to disruption of the natural lands, high urbanization, air, soil and water pollution, deforestation, increased concern for wild animals due to the accessibility of tourism areas.

The Australian model is implemented in Australia, the United States, Canada. This model is based on the concept of preserving and maintaining the natural conditions. It is consistent with public policy and Australia's geographical and environmental conditions: extensive forest areas, multi-species of flora and fauna have been preserved here, and more than one thousand specially protected the natural areas have been created. Ecotourism in this way is aimed at knowing nature and protecting it. Rest in nature in this case is combined with cognitive, sports and aesthetic goals (Ceballos-Lascuráin 1996).

In terms of the tourism industry management in world practice, specialists highlight three models.

The first model provides for a strong and authoritarian ministry that regulates the whole industry. Its implementation requires large financial investments in tourism industry and infrastructure, as well as in the organization of national tourist products' promotion abroad. This model of governance is common in Turkey, Egypt, Tunisia, Thailand.

The second model prevails in European countries. The essence of this is that tourism development is addressed at the level of a multi sectoral ministry, most often with an economic bias. Such ministries usually work in two directions: solve problems of state regulation (regulatory and legal support, international cooperation in the field of tourism, etc.) and carry marketing activities, participate in exhibitions, manage missions abroad. This model of tourism state regulation is common in France, Spain, Italy, Great Britain, which, according to statistics, accounts for about a third of international tourist arrivals (ITA).

The third model, which can be identified when studying the peculiarities of tourism management organization abroad, implies the absence of a central state administration, and all issues are solved on the basis of market self-organization. This model is typical for highly developed market economies (Zhukova 2004).

It should be noted that those countries of the world that have taken the path of state support for tourism, today have a significant export–import item of trade balance (France, Spain, Italy, Greece, Turkey, Egypt, etc.). Legislation has a special role to play, which in these countries determines in detail the relationship between producers and consumers of tourism services, which is significant for industry's stable development.

For example, in France, tourism regulation is the responsibility of the Ministry on transport and public works, which has a Secretary of tourism state. In addition, they are engaged in organizing international cooperation in this industry.

Spain also refers to the second model, where the country plays an active role in tourism industry development. Thanks to a wealth of recreational resources and skilful public policies, Spain has become one of the world's largest tourist centres over the past 60 years.

In the UK, the tourism sector is run by the Ministry of culture, spectacle and sport, which is under the authority directly overseeing tourism as the British Tourist Authority (BTA). In its organizational and legal form—it is a private-entrepreneurial institution. It promotes the country to foreign markets, organizes domestic tourism, and advises the government and other entities on tourism issues. Advertising companies abroad are carried out through a network of representative offices, thanks to media, organization of international conferences and seminars, as well as provision of consulting and marketing services.

About the same way is organized the management of tourism industry in other countries of Europe, such as Germany, Austria, Hungary, Norway, Finland. This is largely due to their presence in the European Union, which dictates legislation harmony of these countries with standards of Europe, including the field of tourism.

An example of third model management is the United States. Since 1961, when the first National Tourism Act was passed, there was a Travel and tourism administration (TTA) within the United States Federal Department of Commerce, which addressed main challenges in the tourism management field: coordination of public tourism policies for the United States benefit; studying of market condition; maintaining statistics of tourism; development of tourism betterment programs, etc. (Bessonova 2016).

Next, we will consider the geography of ecological tourism development and organization model.

In the United States, a special methodology for "interpreting" nature has been developed to improve visitors' understanding concerning landscape aesthetics. Due to its help, the service of reserve or national park translates to public the "language" of park, it's the natural and cultural environment. Interpretation helps people to better understand nature, to engage in interaction with it. From interpretation to understanding, from understanding to high appreciation, from assessment to protection.

The African tourism macro-region offers modern ecotourists huge territories, diverse landscapes, the richest animal and plant world, diverse population. Kenya is a striking example of ecotourism as the most important sector of economy. Since January 2018, the development of tourism and wildlife in Kenya has been the responsibility of one authority, the Ministry of Tourism and Wildlife. The country has a national long-term tourism development programme (Kenya National Tourism Blueprint 2030), according to which African safari is one of the Kenya's two main tourism products. The government of that state has pursued an active policy in ecotourism. Representatives of Kenya's tourism business are taking a specially designed course on ecotourism. The programme was prepared by the Kenya Wildlife Conservation Service. Thus, development of ecological tourism takes place through the use of the natural capital. To date, the natural environment has not undergone significant anthropogenic changes. This means that ecotourism can develop fully in this country.

Ecotourism gained significant development in the Republic of South Africa (RSA). In 1994, the Republic joined the World Tourism Organization. This year was a year of fundamental measures in tourism field. Tourism Council of South Africa provides special training for tourist guides in connection with the wide variety of the natural conditions in country, from desert to humid rainforest. Currently, ecotourism has become the most profitable and rapidly developing type of tourism in South Africa. Many national parks of this country are protected by UNESCO (Drozdov 2005).

In the North American region, ecotourism is represented by a deployed system of specially protected the natural areas. One of the most advanced systems is in Canada, with about 40% of the area being protected. Canada's low population density and vast untouched spaces are favourable for ecological tourism development. The number of tourists wishing to visit these places is constantly increasing, so the recreational infrastructure is being improved: hotels, mountain huts, ski routes are being built; and the track-pedestrian system is developing.

The US, which sends the world's largest number of ecotourists abroad, also has a huge domestic ecotourism market. Protected the natural areas of the United States, which occupy more than 10% of the country's area, are becoming increasingly popular among both residents and foreign tourists. Ecological tourism is not only a means of protecting nature, but also contributes to improving the infrastructure of settlements near environmental routes. In the USA and other developed countries, with little importance of ecotourism at the state level, its role at regional and local levels is constantly increasing (Kosolapov 2005).

Central and South America has the largest variety of flora and fauna species on the Earth.

In France, Great Britain, Holland, Ireland, Germany and Spain, involvement in green tourism activities is encouraged at the national level. Rural recreation in these countries is now the second most popular holiday after sea holidays.

The UK is developing predominantly domestic ecotourism. Most of the international travel directed to the country involves business, educational and cultural tourism. Ecotourism in the UK is an additional means of economic development for local population and land restoration.

In Switzerland, it has historically developed that beginning from the XIXth century, the foreign aristocracy, mostly from Great Britain, rested on picturesque shores of Switzerland's numerous lakes. Thomas Cook's company first organized the excursion in 1863 to Lake Geneva in this country.

International tourism has an important place in Switzerland's economy. Tourism in the country is actively explored. For example, as early as the Second World War, major educational and research centres such as the Institute for tourism research of the Bern University and the St. Gallen school of economics were founded. In addition, tourism development is supported by governmental and social organizations of the country, as it is the only economic alternative to agriculture in mountain areas. However, it also creates problems for country leadership, as it adversely affects the sensitive environment of the Alps and causes inconvenience to local population.

Finland rightly ranks among the most environmentally friendly countries in Europe. This is facilitated by state policy, in which careful treatment of the natural resources and minimal human influence on ecological environment are elevated to the rank of national idea. Ecological tourism is a kind of business card in country.

Values promoted by ecotourism in Norway are highly revered. In 2014, the Norwegian Tourism Council announced the active development of ecotourism in country. Most of Norway's tourist sites possess environmental certificates, among which most of them are owned by Ecotourism Norway and Green travel.

The Asian mainland is fascinating for ecological tourism. India, Nepal are particularly distinguished in the South Asia; China, Malaysia and Japan in the East Asia; Thailand, Indonesia and Vietnam in the South-East Asia.

Australia is one of the world's centres for ecotourism. For about 3000 national parks and reserves have been created here.

On the island of Tasmania, about 30% of territories are on the World Heritage List. Tourists are attracted here by absolutely wild mountainous terrain.

On territory of the CIS, ecotourism development was seriously engaged by many tour operators. Russia pays more attention to the organization and growth of ecological tourism than other CIS countries. The country has significant potential for ecological tourism development. The 45% of Russia's territory has had little human impact, while in the United States such territories are the 4%, in Western Europe—only 2–3% (Gorshkov et al. 1995).

In Russia, local authorities have recently been increasingly concerned with the sustainable development of territories and, consequently, with ecological tourism related to sustainable development of tourism industry. Russia has high hopes for successful development of ecotourism in country, which will become a means of sustainable development. The main vector of ecotourism policy is aimed at the development of cognitive tourism, which according to the Concept of SPNA development in Russia is defined as one of the ecological tourism special types.

Developing countries are priority ecotourism dislocations.

In foreign countries, the development of rural tourism has substantial state support. For example, the State Tourism Administration of China stated that "main focus of China's tourism industry will be rural tourism". Many countries have legislative benefits. In Poland, farmers employed in green tourism are not covered by the law regulating economic (entrepreneurial) activities. These services are not subject to value added tax. The legislation also provides, under certain conditions, an exemption from the income tax of individuals. In Hungarian legislation, "green-tourist" enterprises are not subject to the normative acts that regulate relations in the sphere of entrepreneurship.

Active role of the state contributes to ecotourism development. The example of various countries where ecotourism is developing most successfully shows that sound public policies and regulatory frameworks avoid the shortcomings of nature-oriented tourism, make it sustainable and contribute to regional economic prosperity, while addressing environmental concerns.

Ecotourism management includes two economic models which are revenue model and return model. Revenue model can be used to solve the external problem of scenic spot and the return model can play an important role in coordination among the tourism enterprises based on the economic benefits.

In terms of the ecological tourism the geography, it differs from geography of traditional tourism. Main international tourism flows directed from developed to developing countries. At the same time, however, it is difficult to argue that in the latter, share of ecological tourism in tourism industry is higher, as in countries of Europe and North America domestic ecotourism is widespread. The main factor determining the peculiarities of ecotourism geography is the presence of attractive little-changed the natural complexes. Numerous ecotourists attract territories with maximum biodiversity—equatorial and tropical forests, savanna and mountainous areas.

Difference in sustainable development problems determines the geographical specificity of ecological tourism role in its achievement.

Geographical differences in demand and supply on ecotourism lead to differences in organization and characteristics. Demand depends largely on socio-economic factors (growth of population income, amount of free time, necessity of rest in nature), most of it is concentrated in developed countries. Supply, along with socio-economic factors, is determined by the natural factors and concentrated in both developed and developing countries. Thus, domestic and international ecotourism can be distinguished. Domestic and international ecotourists often pursue different goals. Developed and developing countries also have different objectives in increase of ecological tourism.

Developing countries have an undeniable advantage in diversity and extent of ecological nature. For third world countries, the development of ecotourism is an obvious way out of their "classic impasse".

A large proportion of services provided by countries in the second group are in the hands of local companies, and ecotourism is treated equally a means of economic development and nature conservation.

The third group is characterized by a combination of domestic and international ecotourism. In the United States, due to the huge volume of demand, there is a greater development of domestic ecotourism, while in Canada, Scandinavia, and the Mediterranean region, international ecotourism is taking precedence.

In countries of the fourth group, as in previous one, the vast majority of service delivery is concentrated in hands of local organizations. Some of the functions on organizing ecotourism are entrusted to the state, represented by local governments, and public organizations.

Countries of so-called transition economy in their present form are difficult to identify as any of the listed groups, because of their almost complete exclusion from the world ecotourism market, and their transition condition. Russia and Kazakhstan have significant potential for ecological tourism development among them. In case of successful implementation on planned changes and achievement of high standard for life well-being in the future, Russia and Kazakhstan can be assigned to the third group, and countries of Central and Eastern Europe—to the fourth group.

In international nature-oriented tourism, the main regions generating demand are Western Europe and North America.

Territories most visited by ecotourists are in developing countries of Latin America, Asia, Africa, Australia and Oceania.

Geographical location of country affects the prevalence of tourists from certain countries and regions. For example, the Caribbean, Latin America and the Pacific are dominated by tourists from North America, Africa and South Asia by Europeans. In many former colonies, a large proportion of tourists are subjects of their former metropolitan areas. For example, Kenya is dominated by UK tourists, and they account for a large share in Australia, India, New Zealand. The Madagascar and Mauritius are dominated by tourists from France.

In developing countries, the main regions of international nature-oriented and ecotourism are Central America, north and east South America, South and East Africa, South and South-East Asia, Australia and Oceania.

In Latin America and the Caribbean, ecotourism is the main tourism sector in Costa Rica, Ecuador, Belize, and an important tourism sector in Mexico, Venezuela, Peru and Caribbean island countries.

Brazil had great potential, but ecotourism was not given sufficient attention to the interests of traditional development, and Amazonian herring was being actively cut down, resulting in the loss of valuable the natural and ethnographic resources. The Andean region is promising for ecotourism development.

In Africa, ecotourism is most developed in Kenya, Tanzania, and South Africa. It is also active in Botswana, Madagascar, Mauritius, Namibia and Cameroon. Such countries as Zaire, Ethiopia and other countries have good opportunities (Weaver 1998).

The Nepal stands out in Asia, whose popularity has soared in recent years. Ecotourism is essential in Thailand, Indonesia, Sri Lanka, where exotic nature is successfully combined with interesting cultural and ethnographic objects. Ecotourism in India can play an important role. However, rapid population growth is constantly reinforced man-made press on the country's the natural resources, leading to their deterioration and loss as ecotourism resources. In addition, the Philippines, Malaysia, Bhutan and China can be distinguished.

Australia and New Zealand have long been popular with tourists due to their unique nature. The potential for ecotourism in many Pacific islands is enormous.

The popularity of Antarctica grows rapidly. However, a lack of proper control related to its neutral status can lead to devastating consequences for wounded Antarctic landscapes (Dmitruk 2004).

In addition to international, the domestic ecotourism also is popular in various countries. It is the most developed in North America and Europe (especially the United States, Germany and the United Kingdom). Furthermore, domestic market demand for ecotourism is increasing in some South-East Asian countries (Thailand, Indonesia). Other countries with large volumes of domestic ecotourism include Canada and France.

Emergence of ecotourism projects in the Alpine-Mediterranean region of Europe, along with traditional tourism, is noteworthy. This is facilitated by the presence of the natural and poorly modified environmental areas in some places. However, in the Mediterranean, ecotourism is not easy to compete with beach recreation and other types of tourism (Dedovskih et al. 2002).

Ecotourism concepts in many countries contain an educational component, especially through studying programs for local people interested in ecotourism development. Local enterprises, communities and visitors are influenced through education and intervention, behavioural change and environmental responsibility.

Experience of different countries in ecotourism development shows country-specific and common approaches in formulation and implementation of ecotourist policies harmonized with international recommendations and objectives to create conditions for sustainable tourism development in transition context to green economy. Various approaches, concepts, models and tools are being developed in different countries in accordance with their national situations and priorities. As part of the ongoing debate on the possibility/impossibility of achieving a "better" form of tourism consistent with the principles of global concept on sustainable development, case studies from different countries demonstrate how ecotourism can contribute to the achievement of sustainable development goals.

In general, the growth and diversification of demand is leading to an expansion of ecotourism geography.

Ecotourism today is a dynamic direction, the form of activity in the world tourism industry. But at present stage of development, it is necessary to revise the conceptual understanding of ecotourism phenomenon. The prospect of tourism development, greening all its forms and species is the inevitable future of industry.

References

Aigistova OV, Zabayev YV, Seselkin AI (1996) Vvedenie v biznes turopereitinga (Introduction to the business of tour booking). RMAT, Moscow, p 104
Alexandrova A (1996) Economika I territorialnaya organizatsiya mezhdunarodnogo tourisma (Economics and territorial organization of international tourism). MGU, Moscow, p 105
Almukhamedova OA (2013) Rol' ekoturizma v ustoychivom razvitii rekreatsii (The role of ecotourism in sustainable development of recreation). News South Federal Univ Tech Sci 6(143):8–16
Babkin AV (2008) Spetsial'nyye vidy turizma (Special types of tourism). Rostov-na-Donu, p 252
Bessonova GB (2016) Organizatsionnyye modeli gosudarstvennogo regulirovaniya turizma za rubezhom (Organizational models of state regulation of tourism abroad). Bull Saratov State Soc Econ Univ 1:21–24
Bhuiyan AH, Siwar C, Ismail SM, Islam R (2012) The role of ecotourism for sustainable development in East Coast Economic Region (ECER). OIDA Int J Sustain Dev Malays
Boreyko V (1996) Krasota prirody i zapovednyye ob"yekty (Beauty of nature and protected objects). Bull Wildl Conserv 11:22–25
Ceballos-Lascuráin H (1996) Tourism, ecotourism and protected areas: the state of nature-based tourism around the world and guidelines for its development. IUCN Publications, Cambridge, p 301
Chizhova VP, Dobrov AV, Zakhlebny AN (1989) Uchebnyye tropy prirody (Educational nature trails), p 159
Crouch GI, Brent Ritchie JR (1999) Tourism, competitiveness, and social prosperity. J Bus Res 44(3):137–152
Dedovskih EY, Drozdov AV, Moraleva NV (2002) Ekologicheski turizm kak sovremennaya ideologiya puteshestviya na prirodu (Ecological tourism as a modern ideology of travel to nature). Grif and K, Tula, p 284
Dmitruk OY (2004) Ekologicheski turizm (Ecological tourism). Alterpres, Moscow, p 192
Drozdov AV (2005) Osnovy ekologicheskogo turizma (The basics of ecological tourism). Gardariki, Moscow, p 271
Dunlap RE (1993) From environmental to ecological problems. New York, pp 707–738
Durovich AP, Kabushkin NI, Sergeeva TM (2003) Organizatsiya tourism (Tourism organization). Textbook. Minsk, p 632
Gorshkov VG, Donchenko VK, Kondratyev KY, Losev KS (1995) Ekologiya, ekonomika, politika (Ecology, economics, politics). Zelenyy Mir 31:10
Green Travel Trends 2020: Gen Z and Beyond (2019) Criteo travel study, p 1
Hall CM (2005) Rethinking the social science of mobility. Pearson Education, Harlow, p 347
Hammer T, Mose I, Siegrist D, Weixlbaumer N (2007) Protected areas and regional development in Europe: towards a new model for 21 century. ReseacrhGate, p 272
Information Resource Centre of Ecotourism in Kazakhstan (2010) Ecotourism Information Resource Centre, Almaty
Inglehart R (1990) Culture shift in advanced industrial society. Princeton University Press, Princeton, NJ, p 484
Ivanov AN, Chizhova VP (2003) Okhranyaemye prirodnye territorii (Protected the natural territories). Moscow University, Moscow, p 119
Johnston RJ (1991) Geography and geographers: Anglo-American human geography since 1945. Edward Arnold, London, p 9
Kairova SG, Yesimova DD, Malikova FM (2018) Ustoychivyy ekologicheskiy turizm v Kazakhstane (Sustainable ecological tourism in Kazakhstan). Sci Tour Interact Strateg 8:22–28
Khrabovchenko VV (2004) Ecological tourism (Ecological tourism). Finance and statistics, Moscow, p 208
Kiper T (2013) Role of ecotourism in sustainable development. In: Advances in landscape architecture. IntechOpen, p 31

Kolbovskiy EY (2006) Izuchayem prirodu v gorode (Studying nature in the city). Yaroslavl, p 256
Kosolapov AB (2005) Teoriya i praktika ekologicheskogo turizma (Theory and practice of ecological tourism). KNORUS, Moscow, p 240
Kosolapov AB (2009) Geografiya rossiyskogo vnutrennego turizma (Geography of Russian domestic tourism). KNORUS, Moscow, p 69
Kudakayeva AG (2014) Osobennosti razvitiya ekologicheskogo turizma Rossii (Features of the development of ecological tourism in Russia). Molodoy Uchenyy 8:362–365
Kuserova LI, Kovshov VP (2010) Tourism kak obiect issledovanye v istory geografichesky nauki (Tourism as an object of research in the history of geographical science). Curr Probl Geogr Geoecol 1(7):32
Kuskov AS, Golubeva VL, Odintsova TN (2005) Recreathionnaya geografiya (Recreational geography). Moscow, p 496
Lindberg K (1991) Policies for maximizing nature tourism ecological and economic benefits. Washington, p 44.
Mihalič T (2000) Environmental management of a tourist destination: a factor of tourism competitiveness. Tour Manage 21(1):65–78
Mozgovaya OS (2004) instumenty upravleniye razvitiem ecologicheskogo turizma v regionakh mira (Tools for managing ecological tourism development in regions of the world). Belarus J Int Law Int Relat 2:6
Naibolee seryoznye sosialnye problemy, volnuyushie lyudei (The most serious social problems that concern people) (1998) In: Bulletin of the information network on sustainable development in Central Asia. Special issue dedicated to the forum of environmental NGOs of Kazakhstan, Almaty, p 1
Niyazbayeva A (2006) Osnovnyye napravleniya razvitiya turistskoy otrasli v Respublike Kazakhstan (The main directions of development of the tourism industry in the Republic of Kazakhstan). Al'-Pari 3(4):71–73
Novikov VS (2013a) Organizatsiya turistkoi deyatelnosti (Organization of tourism): a textbook for students in higher education institutions. «Academy» Publishing Centre Bachelor Series, Moscow, p 54
Novikov VS (2013b) Organizatsiya turistkoi deyatelnosti (Organization of tourism activities): textbook for students of higher education institutions. «Academy» Publishing Centre Bachelor Degree Series, Moscow, pp 46–48
Panov IN (1998) Ecologychesky tourism I ego rol v ustoishivom razvitii territorii (Ecological tourism and its role in the sustainable development of territories). Moscow, pp 13–18
Pemecky T (2010) The being of tourism. J Tour Peace Res 1(1):2
Pisarevskaya OI (2014) Osnovy tourism (Tourism basics). Mosow, Feder adepsu for tourism, p 374
Plemarom A (1999) Zapovedi otvetstvennogo turista (Commandments of the responsible tourist). In: Third world resurgence publishing centre bachelor degree series no. 103. Moscow, pp 46–48
Santarem F, Campos JC, Pereira P, Saarinen J, Brito JC (2018) Using multivariate statistics to assess ecotourism potential of water bodies: a case-study in Mauritania. Tour Manage 67:34–46
Shmitz A (2008) Development kurortnoy nedvizhimosti (Resort real estate development). Bilding-KhK Publication, Moscow, p 479
Shtyurmer YA (1978) O problemakh tourisma I zashity ocruzhayushei sredy (On the problems of tourism and environmental protection). Moscow, p 95
Spasibenko LD (2015) Kharakteristika rynka turisticheskikh uslug v Rossii (Characteristics of the tourist services market in Russia). Int Sci-Res J 10(41) Part 1:71–73
Stanev P, Enev E, Atanasova E (1984) Ekonomika i organizatsiya mezhdunarodnogo turizma (Economics and organization of international tourism). Economics, Moscow, p 167
Strasdas W (2002) The ecotourism training manual for protected area managers. German Foundation for International Development, Centre for Food, Rural Development and the Environment, Zschortau, Germany, p 58
The International Ecotourism Society (2019) TIES, p 1

Tourism: Study manual for students of pedagogical institutes on specialties №2114 "Physical education" and №2115 "Basic soldier training and physical education" (1988) Moscow, p 224
Turizm I dovkillya: Serija «Pitannya ta vidpovidi», №3 (Tourism and environment: Series "Questions and Answers", No. 3) (1999) Council of Europe and Citizens' Association "Green Ukraine" Society of the National Environmental Centre of Ukraine, pp 17–29
Volsky VV (1998) Sosialno-ekonomicheskaya geografiya zarubezhnogo mira (Socio-economic geography of the foreign world). KRON-PRESS, Moscow, pp 272–283
Vorobyev VV, Ryashenka SV (1988) Rekreasiya i ohranyaemye territorii (Recreation and protected areas). Irkutsk, p 128
Vsemirnaya konferensiya Turizm i energiya budushego—Sokrashenie vybrosov CO_2 (World conference on tourism and future energy—reduction in CO_2 emissions) (2017) In: EXPO 2017, Astana, Kazakhstan, p 108
Weaver DB (1998) Ecotourism in the less developed world. CAB International, pp 109–134
Worner G (1992) Ekologicheskaya antologiya: Ekologicheskie proizvedeniya zapadnyh avtorov. Sovetsko-amerikanskaya gumanisticheskaya inisiativa (Ecological anthology: ecological works of western authors. Soviet-American humanist initiative). Journal of Moscow, Moscow, p 267
WTO (2001) Tourism 2020 vision, vol 7 global forecast and profiles of market segments. UNWTO, p 139
WTO (2012) Compendium of best practices and recommendations for ecotourism in Asia and the Pacific. UNWTO, Madrid, p 136
Yerdavletov SR (2000) Geografiya tourisma: istory, teoriya, metody, practica (Geography of tourism: history, theory, methods, practice). Almaty, p 336
Yerdavletov SR, Aliyeva ZN (2013) Elektronnye tekstovyye dannyye (Electronic text data). Kazakh National University named after al-Farabi, Almaty, p 273
Zhukov PV (2016) Ecotourism kak podsystema sotsialnogo institute tourisma (Ecotourism as a subsystem of the social tourism institute). Sci Results Technol Bus Serv Ser 1(7):8–11
Zhukova MA (2004) Menedjment organizasy (Management of the organization). Moscow, p 200

The Natural and Recreational Potential of Kazakhstan for the Ecological Tourism Development

2

2.1 General Characteristics of the Natural and Recreational Resources

In general, the nature of the republic is diverse; there are many the natural complexes that are well-preserved in the natural state, which is of great importance for the ecological tourism development.

Tourist and recreational resources combine the natural and anthropogenic (cultural) landscapes, brought to the level of adaptability intoxicated of social needs and possibility of their direct use in tourist and excursion services.

The natural and recreational resources are the natural geosystems, the natural phenomena that have comfortable properties for recreational activities, and can be used for some time to organize recreation and health improvement of certain people contingent.

Anthropogenic or otherwise socio-economic recreational resources are cultural and historical objects (monuments and memorial sites, museums, etc.) and phenomena (ethnographic, political, industrial, etc.) (Yerdavletov 2010).

Recreational Assessment of the Relief

Ecological tourism, as a type of tourism, has more the natural orientation. Ecotourists take into account the features of landscape and climate, richness and diversity of flora and fauna, i.e. the natural resources play a huge role in the choice of recreation areas to rest and travel.

The surface of nature is equally important according to varying territory. Relief, which difference is expressed in terms of relative height of the water flow and degree of subdivision area remains as the leading element of environment because it creates the conditions for diversity of climatic and hydrological phenomena and processes. In addition, variety of relief is a very significant advantage of the landscape. Within Kazakhstan, all elevation levels are observed from lowlands at

K. Iskakova et al., *Ecological Tourism in the Republic of Kazakhstan*,
https://doi.org/10.1007/978-3-030-77462-2_2

absolute altitudes less than 200 m, some of which have negative surface marks, to highlands with up-to-date glaciers.

The total surface increase is from the western and northern regions of Kazakhstan to the southeast and east, the difference in altitudes exceed 5000 m (Fig. 2.1).

If we compare the extreme in terms of altitude values on the point—the Karagiye depression on the Mangystau peninsula, deepened 132 m below the ocean level, with the Khan Tengri peak at the border of Kazakhstan, Kyrgyzstan and China, with a mark of 6,995 m, the amplitude will be 7127 m.

Plains occupy more than half of the Kazakhstan area. From the point of view of ecological tourism, there are interesting peneplains that cover significant rounded hill tops of Saryarka. Peneplains are called (from the Latin pene—almost) almost flat, sometimes weakly hilly surface, which was formed on the site of ancient mountains. The western half of Central Kazakhstan is less elevated, with wider spread of plains (250–400 m), usually weakly dissected. In the east of country the peneplain is more elevated (350–600 m) and intensively dissected as a result of the impact concerning erosion new stage. Isolated low mountains of Shyngystau (height up to 1305 m), Kyzyltas (1565 m), Karkaraly (1403 m), Bayanauyl (1022 m), Ulytau (1133 m), Kokshetau (947 m), Bektauata (1214 m), Shunak (1111 m), etc. All of them abound with original forms of weathering, and in

Fig. 2.1 Khan-Tengri peak (*Resource* Available at https://Kazakhstan.travel/photobank)

combination with the inner-mountain lakes often form aesthetically attractive landscapes. Lowlands of Saryarka have a tectonic nature and are usually confined to the nuclei of anticlinal structures. They are morphologically well expressed and have a stable contrasting appearance, towering 200–600 m above the framing rounded hill tops or peneplain. Slopes of mountains are usually steep (The national atlas of the Republic of Kazakhstan 2010) (Fig. 2.2).

Mountain ranges of West Kazakhstan—Mugalzhary (up to 657 m) and mountain Mangystau (556 m) in relative heights cannot be compared with the low mountains of Saryarka, and even more so of the Southeast Kazakhstan. However, they are very contrastingly elevated above the flat environment, exceeding it by 200–350 m, that there will be no mistake in attributing them to the class of mountainous terrain. Mugalzhary are ridges, elongated submeridianally, with uneven relative heights and soft outlines of peaks. Mountain Mangystau—three parallel sublatitudinal ridges. The highest central ridge of Karatau is framed by cuesta ridges of Northern and Southern Aktau (The National Atlas of the Republic of Kazakhstan 2010).

Among the steppes of Central Kazakhstan, there are small massifs of the strongest rocks: Kuu, Kent, Karkaraly, Kyzylrai. At these points in the long past geological epochs, molten magma broke through the Earth's crust, forming different forms of relief. The most beautiful of them are the Karkaraly mountains, name of which came from the name of national headgear of Kazakh girls.

The mountains, intermountain and foothill depressions of Southeast Kazakhstan include the third Altai (with the highest point of 4506 m), the ranges: Tarbagatai (2992 m), Sauyr (2930 m), Zhetysu Alatau (4622 m) and Tien-Shan ridges:

Fig. 2.2 The Karagiye depression (*Resource* Available at https://Kazakhstan.travel/photobank)

Teriskey Alatau (more than 4000 m), Ile Alatau (4973 m), Kungei Alatau (4677 m), Kyrgyz Alatau (3820), Ugam (4238 m) and their spurs. Absolute elevations are distinguished by high-altitude (more than 3000 m), medium-mountain (2000–3000 m) and low-mountain (up to 2000 m) relief. Highlands are characterized by sharp peaks, steep slopes and modern glaciation, relict forms of nival processing and fragments of ancient peneplain. The middle mountains adjoin to high-mountain relief or in places it is isolated. It is widely distributed, limited by distinct ledges, and generally differs somewhat smoothed forms, although the relative fragmentation of the gorges sometimes exceeds 1000 m. Low-mountain relief fringes the middle reaches and forms independent massifs. Relatively soft forms of relief are characteristic in the Altai, Zhetysu Alatau and Tien-Shan low mountains (The National Atlas of the Republic of Kazakhstan 2010).

Sharply stand out against the background of the surrounding Kokshetau and Bayanaul mountains.

"Earth, ringing as a horseshoe", as the local akyns call the mountain range Kokshetau (Sinegorje). It is not high. The height of individual hills does not reach 200–300 m. The highest peak of this horseshoe ridge is Sinyuha (1047 m above sea level). From it you can see the majestic panorama of the fairy-tale lake of Borovoe in the green frame of pine forests and amazing granite rocks. Especially beautiful is the rock "Okzhetpes" ("Even the arrow does not reach"), composed of large granite slabs. But the most surprising in its bizarre form is the rock "Zhumbaktas" ("Sphinx"). Rising from the emerald waters of lake Borovoe, it fascinates every tourist. And in many other rocks carved by the winds and rains, the art of nature can be seen. This is the well-known resort Borovoe, in Kazakh it is called Burabay. This is truly a blessed edge of silence, crystal clear air, blue mountains and emerald lakes (Yerdavletov 2015).

Bayanaul mountains in the south-west of the Pavlodar region are no less spectacular. This complex rock massif consisting of separate rocky hill groups in various forms is very attractive in nature "oasis" of the whole North Kazakhstan.

Ecological tourism resources are caves, and the development of which is associated with karst phenomena. For the development of scientific, cognitive, and excursion ecotourism, it is necessary to have rare and attractive forms of relief, primarily forms of morphostructure—tectonic faults, volcanoes, both active and passive, mullets, sea terraces, etc. (Fig. 2.3).

The most striking example of such places is the Ustyurt plateau, located in West Kazakhstan.

One of these places is Ustyurt, a plateau in Kazakhstan and Uzbekistan, located between Mangyshlak and Kara-Bogaz-Gol in the west, the Aral Sea and the Amudarya delta in the east.

Plateau area of about 200 thousand km^2 is a clayey and clayey-gravelly desert. Also there are sections of sandy desert. The Ustyurt plateau occupies vast spaces between the Aral and Caspian Seas and has a characteristic feature: the chink is a steep, inaccessible cliff about 150 m high (the eastern chink facing the Aral reaches 190 m). According to scientists, the Ustyurt plateau is the bottom of dried-up sea that existed in these parts in the early and middle Cenozoic era (21 million years

Fig. 2.3 Okzhetpes (authors)

ago). This is evidenced by impregnations of shells in limestone, as well as iron-manganese concretions like billiard balls scattered throughout the plateau. These balls formed at the bottom of the sea, and then, as more resistant to weathering, were on the surface when the limestone and dolomite surrounding them were blurred with water (Fig. 2.4).

The smooth desert relief, interrupted by Cretaceous deposits in the form of rocks and appearing here and there cracks is similar to the Martian landscape from a Hollywood film, and it seems that by an unbelievable chance you are exploring a different, unfamiliar planet. Particularly beautiful is the plateau of Ustyurt during sunsets and sunrises. Cretaceous rocks during sunsets and dawns present an impressive picture, when white rocks are painted in crimson colours.

The nature in the east and south-east of the republic is various and majestic, where the mountain systems of Altai, Zhetysu Alatau and Tien-Shan stretch almost uninterruptedly. Mountainous areas of Kazakhstan are characterized by great picturesqueness and diversity, which creates rich opportunities for organizing ecological tourism. Among the most scenic the natural landscapes that are of great interest among ecotourists is the canyon of the Sharyn river, the left tributary of the Ile river flowing in the east of region. The canyon abounds in towers, pillars, labyrinths of gorges.

"Architecture" is so fantastic that the canyon is called "Valley of Castles", "Sphinx", "Gorge of Witches", etc. The Sharyn Canyon resembles the well-known Grand Canyon of Colorado in the USA. Therefore, it is also called "mini Colorado". Here there are unique communities of plants and animals. These include the

Fig. 2.4 The Ustyurt plateau (*Resource* Available at https://Kazakhstan.travel/photobank)

Falcon, the Black Stork, various snakes, tau-teke, etc. Today, they are becoming less, as many animals avoid entering the canyon, choosing less "civilized" habitats. Currently, the main threat to the Sharyn canyon is uncontrolled tourism, as a result of which the descent to the river through picturesque valley is gradually trampled down; since recent times, motor vehicles began to move directly into the Sharyn floodplain. "Wild" tourists cut trees for fires, pollute the territory and river. Barberry bushes and ash trees are dying. All this led to a sharp increase in loads on the landscape. If today we do not take appropriate measures to conserve the canyon ecosystems, in particular, by moving to the organization of ecological tourism, then excessive loads can lead to degradation of it's the natural complexes (Aliyeva 2002) (Fig. 2.5).

Furthermore, exotic rocks, traces of catastrophic and spontaneous the natural processes and phenomena, are also in interest from the point of view of ecological tourism. So, in the valley of the Small Almaty river there are beautiful rocks, called "gendarmes"; traces of mudflow are found in valleys of the rivers Small and Big Almaty, Esik, Talgar. The most important landscape monuments of inanimate nature in Kazakhstan are the exotic rocks Okzhetpes, Sphinx, Camel in the Kokshetau mountains, granite remains of Stone head, "Baba-yaga" in Bayanaul mountains, the well-known "Singing dune" in valley of the river Ile, burial of the tertiary fauna "Goose flight" on the Ertis river, burial of the fossil flora and fauna in valley of the Torgay river, the inundated depressions of the Caspian lowland, separate karst caves, on the Mangystau peninsula, et al. (Yerdavletov 2015).

Fig. 2.5 The Sharyn canyon (*Resource* Available at https://Kazakhstan.travel/photobank)

When assessing the terrain for ecological tourism, degree of its dissection (densities, depth, steepness of the slopes) is considered first. These characteristics of the relief must be taken into consideration for the construction of hiking trails and terrain cures.

Recreational climate assessment. One of the important elements of the natural environment, which determines ecological tourism development can be considered as the climate. The most important are air purity, favourable insolation conditions, small fluctuations in temperature and humidity, low frequency of fogs, cloudy days, wind protection, etc.

Climate of Kazakhstan is characterized by a sharp continentality, which is due to its position in the continent interior, a significant distance from the open seas and oceans. However, the large extent of Kazakhstan territory excludes the formation of a homogeneous climate. The geographical position of Kazakhstan predetermines significant differences in climatic conditions of northern and southern, as well as central, western and eastern regions. This heterogeneity is manifested most of all in winter, when Kazakhstan territory is under the influence of the Siberian anticyclone western spur. In summer, in connection with the heating of underlying surface in Kazakhstan territory, following this thermal depression is formed. In the transitional seasons of the year through Kazakhstan territory the southern cyclones can be seen, which cause great variability of the weather.

As far as Kazakhstan is located between 41° and 55° north latitude, the duration of sunshine and magnitude of total solar radiation in the north is much less than in the south. Sunshine duration in the north of Kazakhstan is about 2000 h per year, increasing to almost 3000 h in the south.

Continental climate of Kazakhstan is increasing from west to east and from south to north. During the winter months in northern regions of Kazakhstan there is a cold winter. In some years in the north, frosts reach minus 52 °C, but there may be a thaw —up to plus 5 °C. The highest temperature in the north does not exceed 41 °C, and in the south, plus 49 °C (The National Atlas of the Republic of Kazakhstan 2010).

In annual course of precipitation across the republic territory, depending on the geographical location, there are noticeable differences. In North Kazakhstan, maximum precipitation is observed in summer, and in the south and south-east—in spring. In the northern half of Kazakhstan, about 70–80% of annual amount of precipitation falls in warm period of the year. With the advance to the south, the amount of cold period precipitation increases. In some years in the south, precipitation does not drop out for 2–3 months. In such periods, there are so-called "dry" rains, when raindrops evaporate, not reaching the earth surface. In summer precipitation is often accompanied by thunderstorms.

Summer everywhere continues on average 5 months (May–September), the air temperature in the north sometimes rises to 38 °C, in the south up to 46 °C. Humidity is low, mainly windy-dry weather is established.

The republic territory has a long duration of sunshine hours, especially in the summer, which increases the comfort of rest, creates opportunities for solving a number of tourism problems. The most favourable for the development of winter recreation types is the middle belt of the Zailiysky Alatau. Here in winter, the average height of snow cover, depending on relief and exposure of slope, varies within 30–60 cm. The number of days with a stable snow cover reaches 90–120 per year. There are so-called weather inversions, when on the slopes of mountains warmed by the sun, it is warmer than in the valleys. And the majority of holidayers rush to areas like Shymbulak (Yerdavletov 2015).

In general, the climatic conditions of a republic are considered favourable for the ecological tourism development.

Recreational water resources assessment. Surface waters of the Republic of Kazakhstan include concentrated water reserves in the forms of land surface relief —water objects. These include: water bodies (seas and lakes), marshes, glaciers and watercourses (rivers and equated channels).

Surface water is a necessary part of the resource potential in territory. Surface water is an important component of the natural and resource potential of territory. On the one hand, they perform essential for society environment-forming and life-supporting functions, participating in the process of hydrological cycles concerning the formation of territory the natural conditions (climate, relief, flora and fauna). On the other hand, surface waters are the most used the natural resource for satisfying a wide range of human needs.

From the point of view of ecological tourism, the most interesting are: clean little-studied rivers; rivers where monitoring studies are constantly conducted, rivers—places for feeding and spawning of attractive fish species; watercourses, the nature of their ecosystems under the natural and anthropogenic factors influence, which varies sharply from the source to the mouth; the rivers are unusual from geological and hydrological terms.

The important the natural factors in formation of recreational activities are rivers and lakes. They decorate landscape, create a favourable microclimate, allow holidaymakers to engage in water sports and fishing. Under the recreational water use should be understood the activities of population associated with implementation of various recreational activities (recreation, sports, and tourism) in the water areas and in proximity to water objects.

On the territory of Kazakhstan, there are eight water basins: Aral-Syrdarya, Balkash-Alakol, Ertis, Zhayik-Zhem (Ural-Emba), Yesil, Nura-Sarysu, Shu-Talas and Tobyl-Torgai. Water objects of special national importance are as following the Caspian Sea, lake Balkash, lake Zhaysan, the Alakol lake System, the Ertis river (Fig. 2.6).

In Kazakhstan, due to the arid climate, river network is poorly developed. Even in the northern and mountainous parts of the republic, where there is a lot of precipitation, density of the river network averages only 6–8 km per 100 km^2, and in the deserts—two to three times less. Many rivers of the republic, flowing through semi-desert and desert regions, run out of water on the way, not reaching water bodies.

Some of them, especially the rivers of West, Central and partly South region (Big and Small Uzen, Sarysu, Shu, etc.), are divided into separate stretches, salinized and end with "blind" channels.

On the republic territory there are about 85 thousand rivers and temporary streams, of which more than 8 thousand have a length more than 10 km. Most of Kazakhstan rivers belong to the Caspian Sea basin of the Aral Sea, the Balkash

Fig. 2.6 The Ertis river (photo taken by Artemyev A.)

lakes, Alakol and Teniz. Only the Ertis river belongs to basin of the Arctic Ocean. The highest density of the river network (0.4–1.8 km / km^2) is observed in the high-altitude regions—Altai, Zhetysu Alatau and Ile Alatau, the smallest ones are the areas of the Aral Sea and Caspian Sea sand deserts (less than 0.03 km/km^2). The largest republic rivers are: Ertis, length within Kazakhstan is 1700 km, Syrdarya is 1400 km, Zhaiyk—1082 km.

Distribution of river flow through the territory is extremely uneven. The largest volumes of river flow are formed in the Ertis and Balkash-Alakol basins (up to 73–86% of total resources). In Nura-Sarysu, Esil and Tobyl-Torgai basins in low-water years, local runoff is practically absent. (The National Atlas of the Republic of Kazakhstan 2010).

Particular interest for ecological tourism is the water resources of Kazakhstan mountainous regions. The main water bodies here are the rivers. These rivers are characterized by two maximum flow rates—floods and prolonged summer high water, associated with intensive melting of glaciers. Highland-glacial rivers are quite full and successfully used for rest by the water, picnics, for bathing and tempering, and for watering.

High-mountainous zone (above 3000 m), where the majority of mountain rivers flow is formed, and distinguished by a small set of recreational activity types (mountain climbing, mountain and scientific-cognitive tourism).

River areas of the middle zone (1600–3000 m) are distinguished by considerable water content, wider and located valley sections covered with various vegetation, and, as a consequence, great recreational opportunities. These include a massive rest by the water, tourist trips along rivers and streams, adoption of sun and air baths, amateur fishing, etc. The most full-flowing rivers are used by tourists-water workers for rafting.

The most developed in recreational term are the areas of low-mountain zone rivers. As a rule, rivers at the exit from the mountains are quite abundant, with beaches, overgrown with arboreal and grassy shrub vegetation. Tourism infrastructure such as sanatoriums, holiday houses, houses, children's recreation camps, etc. is well developed here. The territory adjacent to riverbeds is widely used for summer mass recreation (picnics, outdoor games, sports, gathering of wild plants, etc.), and the rivers themselves are used for bathing and taking sun and air baths. For recreational and water management purposes, artificial reservoirs and dams are sometimes used, which are created on mountain rivers (the Shelek river—the Bartogai reservoir, the Big Almaty river—the Sayran reservoir, etc.).

Waterfalls play a major aesthetic role, for example, those located in the river gorges Turgen (Medvezhiy, Kairak, etc.), Gorelnik, Butakovka, Uzinkargal, Big Almaty (Shukur, Ayusai), a series of small waterfalls in the river basins Small Almaty, Talgar, Bayankol, Temirlik.

The largest river of Kazakhstan–Ertis begins in China and flows between Kalbin and Narym mountain ranges. There are many small rivers in Ertis, such as Kurchum, Uba, Ulba, etc. And there are many small rivers: Ak-Berel, Turgysyn, Bilezekty, Berezovka, Kokpekty, Small Ulba, Kalzhir, Kalguti, Ulan, Boken and others.

Plain rivers of Kazakhstan–Ertis, Syrdarya, Ile, Zhaiyk, Zhem, Tobyl, Yesil, etc. provide opportunities for recreation, amateur fishing and hunting, rowing, as well as for alloys of low complexity categories.

Mountain rivers, depending on water content, can be used by tourists for rafting.

In Kazakhstan, there are several thousand lakes with a total area of over 20 thousand km^2. Most of them are in the north and east of the republic, especially a lot on the plains of Tobyl-Esil interfluve and Torgay depression, as well as in the Akmola region and the Ertis valley (Yerdavletov 2015).

In Kazakhstan, there are two main types of lake basins: tectonic and exogenous origin. Large reservoirs, such as the Caspian and Aral Seas, lake Balkash, Teniz, Alakol, Sasykkol, Markakol and others, are located in the tectonic basins. Most of the lakes are in forest-steppe zone and the northern part of steppe zone.

North Kazakhstan accounts for 45% of all lakes, for Central and South—36%, other regions—19%. Under the conditions of water exchange in the republic, drainless lakes are predominating. During the spring snow melt, the level of lakes rises due to the influx of large water amounts. By the middle of summer, thanks to intensive evaporation and filtration, the level of lakes is dropping and some ponds completely dry up. (The National Atlas of the Republic of Kazakhstan 2010).

In desert and semi-desert zones the lakes, as a rule, are salty and bitter-salty, contain sulfates, cooking and Glauber's salt and, of course, are unattractive for tourists. However, some lakes are well-known for therapeutic muds, for example, Teresken, near the Zhanakorgan station and Muyaldy near Pavlodar. Fresh and slightly salty lakes, especially the Caspian and Aral Seas, were formerly rich in fish. Lakes and reservoirs are of interest from the point of view of recreation and tourism. This is a place for organized and unorganized recreation, sports, in particular, sailing and rowing, a place for bathing and taking sun and air baths. Lakes, in addition, represent aesthetic and cognitive value. In recent years, lakes and reservoirs of Kazakhstan, in particular, the Caspian Sea, Kaindy, Kolsai, Bukhtyrma, Alakol, Borovoe, Markakol are beginning to develop divers.

Exclusively attractive objects for ecotourists are mountain lakes and reservoirs. Most often, the water in the mountain lakes is cold and unsuitable for bathing, but nevertheless the surrounding landscape combined with pure water attracts nature lovers.

On the territory of the Tien-Shan, there are about 100 small and medium-sized mountain lakes. The largest of them include the Big Almaty lake, Zhasylkol, Akkol, Kolsai lakes, Kaindy, Bogatyr, Karakol, Tuzkol, and others.

The most interesting for ecological tourism are the tectonic lakes, which include the Big Almaty lake, Kolsai lakes, Issyk, Kaindy and others. The Issyk lake, which was one of the favourite holiday destinations before the catastrophic mudflow in 1963, is a classic example. Approximately 5–10 thousand years ago, the collapse of one of the spurs blocked the ravine of the Issyk river, forming a picturesque lake. As a result of the mudflow in 1963, the lake almost disappeared, but the area did not lose it's the natural and recreational significance.

One of the most visited water bodies of Zailiysky Alatau is the Big Almaty Lake, which is located at an altitude of 2005 m.

Recreational value of the lake is determined by a group of factors: mountain landscape, the reservoir beauty, available roads, accessibility and, finally, the proximity of large city.

The Kolsai lakes located in Kungei Alatau in the altitude range 2000–2500 m have become very popular among tourists. The total area of three lakes is about 1.35 km^2, and the maximum depth of average lake is 85 m. "The Kolsai kolderi" national park is open here.

The largest lakes in the Altai are Zhaysan (5000 km^2) and Markakol (455 km^2). Kalba is the land of beautiful lakes. The most beautiful high mountain lake Markakol, located in the arches of the Kurchum range, which is the largest lake on Altai. Here is the Markakol state reserve. In addition to the natural landscape, a valuable species of fish was taken under protection—lenok (ukuka). The beauty of Rakhman lake is fascinating, and it is located among steep wooded slopes of mountains at an altitude of 1760 m above sea level. Near there are outcrops of thermal waters with a temperature of 22–43 °C plus, in their medicinal properties, close to the waters of Tskhaltubo and Belokurikha.

Present day, there are more than 200 reservoirs in Kazakhstan with a total capacity of more than 95.5 km^3. The largest of them are Bukhtyrma (on the Ertis river), Kapshagay (on the Ile river), Shardara (on the Syrdarya river), High-Tobyl and Karatomar (on the Tobyl river), respectively, Nur-Sultan and Sergeevsk (Fig. 2.7).

Fig. 2.7 Bukhtarma water reservoir (*Resource* Available at https://Kazakhstan.travel/photobank)

On the territory of the North Tien Shan, there is a Bartogai reservoir, built on the river Shelek, undoubtedly of interest. It is located 185 km from Almaty at an altitude of 1010 m. Its capacity does not exceed 320 million m^3, the area of the water mirror is 14 km^2. For recreational purposes it is still not used enough.

Glaciers (their number is not more than 2000) occupy an area of about 1500 km^2 (2004 y.), including in Zhetysu Alatau 700 km^2, in the Ile (Zailiysky) and Kungei Alatau—500 km^2, in Teriskey Alatau—90 km^2 and Kazakhstan Altai and Saura—70 km^2. Currently, glaciers are rapidly shrinking. Glaciers together with the high mountain landscapes surrounding them represent an important the natural and recreational resource for the ecological tourism development.

Hydro mineral resources are one of the main the natural healing factors used in sanatorium and spa practice. On the republic territory they are divided into conventional zones: southern, western, northern, central, eastern.

On the territory of the North Tien Shan, the outcrops of mineral water sources are mainly confined to the Ile Alatau and Ketmen ridge (Uzinkara). According to the chemical composition and mineralization of water, the Zailiysky springs are predominantly sulfate and hydrocarbonate-sulphate-chloride-sodium. Among the most well-known are Alma-Arasan springs, on the basis of which the Alma-Arasan resort can function. Warm hydrogen sulphide source "Gorelnik", located at the confluence of the Gorelnik river in Small Almaty at an altitude of 1700 m, and used by holidayers primarily in the summer.

Kuram sources are located beneath the Bokaydyntau mountains, 95 km east of Almaty. Chemical composition is chloride-sulphate sodium-calcium.

Recreational water resources of the republic create all conditions for the ecological tourism development.

Recreational assessment of flora and fauna. Floristic tourist and recreational resources unite all variety of the flora having medicinal properties, scientific and informative, the natural and aesthetic recreational and medico-biological value and participating in process of restoration concerning personal spiritual and physical forces.

Involvement of floristic resources in the process of tourist and recreational activities can be perceived visually—landscapes, excursion objects.

The value of vegetation cover as a tourist and recreational resource is huge. Forest massifs contribute to an increase in the oxygen content on air and its ionization (the process of ion formation on air that exerts human body cleansing effect).

Forest tracts can also be used for gathering mushrooms, berries and hunting.

Grassy associations (meadows, steppes) for recreation are vital, since, occupying open spaces (glades, fringes), they diversify the landscape.

Desert vegetation is characterized by extreme poverty of species composition, low productivity of biomass and thinning of vegetation cover. Therefore it has no great recreational value. But spring in deserts and arid steppes of Kazakhstan are the most beautiful, fertile time of the year (Fig. 2.8).

Fig. 2.8 Steppe in spring season (own authors)

Ephemerals and ephemerides in the early spring embellish the steppes and deserts, usually monotonous in colours, indeed in all colours of the spectrum. And numerous brightly coloured tulips and poppies give the desert and adjacent steppe and mountain landscapes a special charm.

Agricultural land is an essential element of landscape and does not reduce the recreational significance of the territory. If it is large, then such a territory is not of great interest for recreational development. The alternation of fields with copses and woodlands imparts a unique character to landscape and does not worsen the recreational qualities of this landscape.

When assessing land, the main factors that determine the value of the natural complex are abundance, diversity of plant species, quantity and productivity of biomass, the presence of berries, fungi.

Forest resources in the republic are unevenly distributed, woodland in such areas as Aktobe, Atyrau is only 0.03%, and in East Kazakhstan, North Kazakhstan reaches 14% (Tourism of Kazakhstan 2015).

Distribution of vegetation corresponds to the allocation of soils, reflects the patterns of latitude and altitude zonal and increase in the degree of continental climate. On the plains, latitude zonal is clearly evident in the plant cover.

The following the natural zones are distinguished on the Kazakhstan territory: forest-steppe, steppe, desert, and in the mountain's various high-altitude belts.

Forest-steppe zone. Forests here are represented by numerous pins and consist mainly of hardwoods—birch, aspen and willow. The predominant species are birch. These forests have gone through unsystematic cutting down in former years and therefore have now been thinned out (Fig. 2.9).

Many splints of the forest during this period completely disappeared, as they were cut down completely.

Fig. 2.9 Birch scraps of Ulytau (photo taken by Artemyev A.)

The steppe zone passes through the Kostanay, Kokshetau, Akmola, Pavlodar, and partly East Kazakhstan, Aktobe and West Kazakhstan regions (Fig. 2.10).

Red feathers, lessing, korzhinsk, mixture of sand and sawdust, fescue, and from various grass—steppe sage, sage-leaf mullein, wormwood, estragon dominate in the steppe.

Wormwood-fescue-feather grass communities, sparse and impoverished, appear on solonetzic southern black humus earths. In meadows and saucers with meadow-black humus earth soils, meadow and steppificated meadow communities of ground reed grass, crocheted crooked grass, medicinal herbs are common. On rubble and sandy loamy soils there are sites of shrub steppe with spiree of hunter and pea tree.

In the low-mountain massifs of Kazakh hillocky area in the steppes, there are red feather grass, mixture of sand and sawdust, fescue and wormwood predominate. On stony sites grows Dunce's Caps with fleshy leaves (Tourism of Kazakhstan 2015) (Fig. 2.11).

In the dry steppe subzone, the fescue-feather grass steppes with predominance of feather grass lessing or the feather grass, mixture of sand and sawdust, fescue. In the east there is a mountain Kyrgyz feather grass. In places a significant role is played by a large xerophytic feather grass. There is an abundance of thin desert oat. Depleted xerophytic motley grass is fluffy chest, pyrethrum, wormwood. In the

Fig. 2.10 Feather grass steppe (*Resource* Available at https://Kazakhstan.travel/photobank)

Fig. 2.11 Kazakh hillocky area (photo taken by by Artemyev A.)

spring there are ephemera and ephemeroids (bluegrass bulbous, tulips). By the valleys and beams sometimes there are shrub thickets of pea tree and meadowsweet.

Low-power soils of the hill slopes are covered with steppe vegetation from fescue, wormwood of cold and lessing-like mixture of sand and sawdust. On solonetzic soils there is a wormwood-fescue communities with white wormwood. On steppe solonetzic grow fescue, chaste tree, wormwood Shrenka, pyrethrum, and on cortical and solonetzic—black wormwood. In the depressions, while there is an absence of salinization in meadow and meadow-chestnut soils, grassland meadows are common, also groundless brome, pyrea creeping.

Forests in the steppe are located in separate massifs and azonal and intrasonal types of vegetation. Depending on the soil, hydrographic and orographic conditions of growth in the steppe zone, following types of forest vegetation can be identified (Tourism of Kazakhstan 2015):

(a) steppe birch and aspen scraps of forests;
(b) forests of the sandy-and-coniferous strip in Ertis lake side;
(c) pine forests along granites of hillocky area;
(d) floodplain forests.

Vegetation of desert zone in the South Kazakhstan region depends on the characteristics of the substrate.

On the sands fixed by vegetation, there are a lot of gill, and on dusty ones there is gray wormwood. In these sands, groundwater flows closely, and moisture-loving plants appear in the valleys between the sandy hillocks and ridges-loch (jhida), willow, and reed.

The mountains are located in eastern and southeastern parts of Kazakhstan. There is a clear trace of the high-altitude waist.

In high middle reaches of the Tien Shan, mixed and coniferous forests predominate (Tien Shan fir, Semenov fir, juniper), in Zhongar Alatau—Tien Shan fir, juniper and Siberian fir, also in Saur—Tien Shan fir, Siberian fir and larch tree, in the Altai there are Siberian species (fir, spruce, cedar, forming a black taiga) (Fig. 2.12).

In highlands of the Tien Shan—a zone of high-mountain meadows and meadow steppes, in Zhongar Alatau—mountainous zone, in Tarbagatai and Saur—mountain meadow and zone of bald mountains, in Altai—mountain meadow, mountain tundra and zone of bald mountains. In its peculiar forms (water grass, lithophilic lichens) vegetation penetrates into the glacial-nival zone, which is present in all these mountain systems, except for Tarbagatai.

Spruce forests begin, they reach an absolute height of 2700–2800 m and up to 2900 m, alternating there with scrubland–juniper. Creeping juniper thickets reach the alpine meadows (Tourism of Kazakhstan 2015).

Faunistic tourist and recreational resources are considered as renewable, because they have the capacity for self-recovery, are indispensable and economically viable. They act simultaneously and with a help of technologically obligatory resources, directly participating in the process of restoring spiritual and physical strength of

Fig. 2.12 Tien Shan spruces (own authors)

person, and contributing to recreational process, forming complex tourist and recreational resources. All together, they create a high degree of any landscape attractiveness.

Involvement of wildlife resources in the process of tourist and recreational activities can be different in nature and as an element of attraction perceived visually.

Fauna of Kazakhstan is rich and various. Fauna has 181 species of mammals, 500 species of birds, out of which 396 species are found on nesting, 50 species of reptiles and others. (The National Atlas of the Republic of Kazakhstan 2010).

Fauna of Kazakhstan—formation is geologically young, ice formation played a big role in its formation. The ancient fauna, which had a tropical appearance and many features of similarity to Indian and African, became extinct in glacial times, and individual elements survived only in southern Kazakhstan (boar, tiger). Cold-loving types were installed.

On the territory of North and Central Kazakhstan, the most widespread are representatives of steppe fauna.

Among forest species, hare is inhabited—mountain hare, northern red-backed vole, bullfinch, goldfinch, black woodpecker, chickadee. Earlier there were wood grouse and brown bear. A significant number of the European fauna representatives are pine marten, mink, desman.

In semi-desert the fauna is characterized by an abundance of reptiles. Many lizards: 3 species of toad-headed agamis, reticulate and linear lizard, gundis gecko, and agamids. In the northern parts of the Kyzylkum desert, a large (up to 1 m or more) gray monitor is penetrated from the south. Steppe tortoise lives in small barchans dune depressions.

In mountainous areas there are members of forest fauna, mainly squirrels, sable, chipmank, coachscrew, etc.

In coniferous forests there are lynx, roe deer, marmot, etc.

For high-mountain meadows and meadow-steppes of Tien-Shan, a mountain sheep is characteristic such as argali, among mountain goat are teke, snow leopard, Tien Shan (white-footed) bear, red creeper, and more voles. Among the birds are snowcock, the Alpine jackdaw, red-tailed jackdaw, mountain finches, and high-altitude waltzers. In the subalpine belt of mountain meadow zone there are original flightless wingless locusts (Tourism of Kazakhstan 2015).

The great variety of flora and fauna of the Republic of Kazakhstan creates all conditions for ecological tourism development.

In the vast territory of republic, where you can find almost any geographical area, any vegetable community, there are many amazing places in beauty that are ecological tourism objects. Each of the five Kazakhstan regions is interesting and attractive from the point of view of ecotourism.

2.2 Regional Features of Kazakhstan's the Natural and Recreational Resources

West Kazakhstan. West Kazakhstan, which includes Aktobe, Atyrau, West Kazakhstan and Mangistau regions, is the first largest territory in the country (728.5 thousand km^2). West Kazakhstan also includes the northeast coast of the Caspian Sea, the largest lake of the Earth planet.

West Kazakhstan is the largest oil and gas industry in the country, there is also mining of chromites and nickel ores, production of oil and gas equipment, ferroalloys, mineral fertilizers, covering materials. This is the main area of fishing and fish processing.

The natural conditions of the region are more severe than in other regions of Kazakhstan, which is explained by domination of desert and semi-desert zones on its territory. And yet, there are the natural objects in West Kazakhstan that attract tourists—the Ural river with its coastal thickets, lake Shalkar, the Caspian Sea, Ustyurt cliffs, etc. On the West Kazakhstan territory there is the Ustyurt reserve, 7 nature reserves, 2 protected areas (northern part of the Caspian Sea and Kenderly-Kayasan), 2 nature reserves (Irgiz-Turgai and Akzhayik) and 1 botanical garden-Mangyshlak.

The most notable objects are caves of Mangistau and the Ustyurt plateau, the Ustyurt cliffs, the deepest depression in the CIS—Karagiye, lake Shalkar, Aktogai plateau, Tamshaly, Aktau mountains, and finally, the Caspian Sea.

The cliffs contain clayey-calcareous rocks of tertiary age forming the Ustyurt plateau; beneath steep cliffs there are outcrops of underground waters; and reed meadows grow there.

Karagie—the deepest in the CIS dry depression, located at an altitude of 132 m below sea level. It is located in the desert zone centre on Mangystau peninsula. The depression is bordered by steep slopes, and at the bottom there is a drying lake and salt marsh. The valley's formation is associated with the efflorescence processes on salt-bearing rocks, with subsidence and karst processes that took place on the Caspian Sea coast.

Aktolagay plateau. Aktolagay is a unique terrain. You can meet paleontological monuments of the Mesozoic era—vertebrae and bones of dinosaurs, fossil beetles, shellfish, scales of huge unknown fish, teeth of sharks, which lived here millions of years ago. During the Cretaceous period, this area was the bottom of shallow sea, but sometimes, the sea retreated, and areas of land that were eventually covered with vegetation were exposed. Here you can find fossil trees. World of the Cretaceous sea is rich.

Tamshaly is a remarkable oasis of the desert. You can reach this place by walking through the canyon or moving along a cliff parallel to its brute walls, consisting of multicoloured layers of rocky layers. On the eastern edge of Atyrau region, the vast steppe spaces find their edges, and perfectly flat horizon line is replaced by sharp peaks of bizarre Cretaceous rocks. Once here was the bottom of world's ocean. Shark teeth, dinosaur remains, prints on the stones of unseen plants and unknown liveness can all still be found on the Cretaceous plateau. Cretaceous' mountains are found in different regions of the Earth. We also have these amazing witnesses to the ancient history of planet in Kazakhstan. White rocks emotionally affect people that its historical atmosphere of the natural charm attracts people.

Finally, the Caspian Sea with bitter salt water. On its banks there were large trade routes.

On Mangistau's peninsula located the deepest valleys in Kazakhstan and one of the deepest valley on the globe—Karagie with a mark of 132 m below the world ocean level. Thus, all available facilities of West Kazakhstan contribute to ecological tourism development.

Central Kazakhstan. The administrative area of Central Kazakhstan includes one area—Karaganda, but it is the largest among the country regions (428 thousand km^2). Central Kazakhstan is located in the middle part of republic, in small river basins of Nura and Sarysu, within the Kazakh folded country—Saryarka.

This is a large industrial area, the coal mining, ferrous and non-ferrous metallurgy, metal-intensive engineering and chemical industry are particularly well-developed here. Considering the development of virgin and fallow lands, the importance of region also increased significantly in the agricultural products production, that is grain, meat, milk.

On the territory of Central Kazakhstan there are 2 national the natural parks—Karkaraly and Buiratau, 9 nature reserves and 1 botanical garden—Zhezkazgan.

Among the steppes there are small massifs of the strongest rocks: Kuu, Kent, Karkaraly, Kyzylrai. At these points in the bygone geological epochs, molten magma broke through the earth's crust, forming various forms of relief. The most beautiful of them are the Karkaraly mountains, whose name comes from the name of the national headdress of Kazakh girls (Fig. 2.13).

Streams flowing through the gorges are fed by the sources. Pashennoe, Bassein, Shaitankol small lakes are charming with their beauty. Mountains attract to themselves with fabulous beauty, purity of mountain air, filled with the aroma of pine forest. The tops of Karkaraly mountains do not go beyond the clouds, their height is 1400 m above sea level. But they are famed for their uniqueness. Birches growing on rocks and cliffs, talkative transparent streams, cutting pine forests, clean springs, and lakes. There are many mushrooms and berries. Strawberry ripens on open sunny fields; also in the shady places grow currants and raspberries.

Lake Karasor (or Tuzdykol) has an area of about 250 square kilometres. The lake is rich in fish. Content of salts in the Karasor water exceeds the salt content in seawater. There is therapeutic mud on its shore. About ten small rivers flow into the lake of Karasor.

Lake Balkhash. There is a legend that the rich man of Balkhash had a beautiful daughter, Ile. The rich man announced the holiday, and winner was promised to have his daughter as a wife. Beautiful Ile and the poor shepherd Karatal loved each other. Ile helped Karatal to win the suitors match. But Balkhash did not want to give his daughter to a poor man. Then Karatal and Ile fled away. An angry father could

Fig. 2.13 Karkaraly mountains (*Resource* Available at https://Kazakhstan.travel/photobank)

not catch up the fugitives. And then he turned them into rivers, and while rushing to meet them turns into a lake, absorbing the disobedient daughter. And the lake has a colour of gray blueness (Fig. 2.14).

Only here the extincted Aral thorn and Aral barbel are survived (Central Kazakhstan. About the regions of Kazakhstan. Tourism of the regions of Kazakhstan https://www.orexca.com/rus/Kazakhstan/regions/central_Kazakhstan.htm).

The Bektauata mountain range. At a height of more than a thousand meters rose pink peak Bektauata is an oasis of the Northern Balkhash. In clear weather, it can be seen for a hundred kilometres, so it is not for nothing that it is called the lighthouse of lake Balkhash. The mountain range of Bektauata is small in area (4000 ha), but it contained an unusual variety of bizarre rocks, thorny ravines (Central Kazakhstan. About the regions of Kazakhstan. Tourism of the regions of Kazakhstan https://www.orexca.com/rus/Kazakhstan/regions/central_Kazakhstan.htm).

Central Kazakhstan has all the necessary conditions for ecological tourism development.

North Kazakhstan. North Kazakhstan includes four areas: Kostanay, North Kazakhstan, Akmola and Pavlodar. This is the third in the territory (6009 thousand km^2). In north, the Republic of Kazakhstan borders with Russia. Its border is located in the river basins of Tobol, Ishim, Ubagan, from southern slopes of the Ural Mountains in the west to the plain steppes of Irtysh in the east. Therefore, the natural conditions of North Kazakhstan have many similarities with the nature of neighbouring Western Siberia. The region of the Kazakhstan northern part is mostly flat; nature alternates its beauty of the steppe open spaces with the "islands" of

Fig. 2.14 Lake Balkhash (authors)

white birches and weeping willows framed by the blue surface of lakes. Climate in the region is quite severe, sharply continental with temperature fluctuations from minus 45 °C in winter to plus 45 °C in the summer (North Kazakhstan. About the regions of Kazakhstan. Tourism of the regions of Kazakhstan https://www.orexca.com/rus/Kazakhstan/regions/northern_Kazakhstan.htm).

First, there are 2 reserves (Nauryzym and Korgalzhyn) and 3 national nature parks (Bayanaul, Burabai, Kokshetau), as well as 12 nature sanctuaries, 21 the natural monuments and 2 nature reserves (Altyn dala and Ertis ormany).

Kokshetau-Sinegorye should be noted especially, i.e. the edge of blue mountains. The place is also known as Borovoe. Borovoe is a picturesque resort area, its travellers call it: "the edge of blue lakes", "the edge of blue mountains", "the pearl of Kazakhstan", "the island of health", "Kazakhstan's Switzerland". And all these names equally indicate the unique beauty of these places (Yerdavletov 2015). There are sanatoriums, rest homes, children's health camps.

On lake Maybalyk there is a spring, the mineral waters of which are superior in quality to many others, known in the former USSR. In the steppe lake Balkamsor is the richest deposits of medicinal mud (Fig. 2.15).

One of the most attractive in the natural respect "oasis" of North Kazakhstan is Bayanaul, a mountain forest oasis located in the Pavlodar region. Pearl of Bayanaul is a beautiful lake called Zhasybay, bearing the legendary hero name. Pine forests

Fig. 2.15 Borovoe (authors)

and bizarre cliffs adorn the lake shores, well-known for its clear and very "soft" water. This is a vast recreation area (Yerdavletov 2015).

There is the greatest interest for ecotourism in the Nauryzym reserve, founded in 1931. Here is the great Torgai bird migratory path. In addition, the soil-climatic border passes here, where the West Siberian lowland passes into the steppe expanses of Kazakhstan.

The nature reserve has unexpected combinations itself, precisely as the place where representatives of the plant and animal world are gathered. Here you can see a bustard and white partridge, a hare-rabbit and fox-corsac, red duck-ataiku and black swan. The reserve is especially interesting in the spring, when there is a massive flight of birds. More than 150 species of different birds can be found here, some of them hibernate in Africa, the Mediterranean, India, Iran, Turkey, and come back in spring (Fig. 2.16).

The main reserve area is the salt lake Tengiz. Korgalzhyn reserve is called the edge of fearless birds. In the spring and autumn here they rest more than three million. During these same periods, numerous herds of nomadic saigas pass through the western part of the reserve. Here the nesting colonies of the rarest bird as flamingos also can be seen. This is the most northerly part in the world, and they only nest in Kazakhstan. The flamingo is a cautious bird, not everybody can see them up-close and personal.

North Kazakhstan has great potential for ecological tourism development.

Fig. 2.16 Tourism objects of the Bayanaul (*Resource* Available at https://Kazakhstan.travel/photobank)

East Kazakhstan. The nature of this region is diverse. In the south-east rises Saur and Tarbagatai, and between them and the Southern Altai, lies the Zaisan hollow. Below Ust-Kamenogorsk to Semipalatinsk, the Irtysh flows along a wide valley (Eastern Kazakhstan. About the regions of Kazakhstan. Tourism of the regions of Kazakhstan https://www.orexca.com/rus/Kazakhstan/regions/eastern_Kazakhstan.htm).

East Kazakhstan occupies the tenth part of the Kazakhstan area. Its territory is the smallest among the economic regions of Kazakhstan. But despite this, it has one of the most beautiful the natural areas of the country, there are two reserves (West Altai and Markakol), one national park (Katon-Karagai), five nature reserves, one nature monument—Sinegorsk fir grove, one botanical garden—Altai and 1 forest nature reserve—Semey Ormany.

East Kazakhstan is interesting and attractive for tourists. Everything on this earth can be found here: barely noticeable trails, wide tracts between cities, twilight gorges and light valleys, cheerful rumble of mines and silence of grain fields. Clean air watered by Altai forests, river coolness, picturesque nature, and attractions appeal people from all ends of the country.

Lake Markakol and Markakol nature reserve attracts particular interest from the natural objects of the East Kazakhstan region, as well as the well-known Rakhmanov—the "Altai Matsesta".

Lake Markakol, located at an altitude of 1485 m is the largest in Altai. Its area is 544 square kilometres, it feeds on many frozen springs and five mountain rivers. Connoisseurs of the Altai mountain call Markakol the crown of beauty in the whole region. Water in the lake is transparent, like a tear, the bottom is clearly visible at a depth of 6–7 m. But it can also become lead-gray, when heavy clouds descend from the mountains. Lake Markakol in beauty is comparable to the deepest lake in the world—Baikal. Anyone who has ever visited Markakol will forever remember the sunny meadows on its bank, the refreshing noise of foliage, the incessant talk of crystal clear streams, the invigorating coolness of snowy peaks. Mountains on the territory of Markakol reserve has absolute heights from 1485 m (lake Markakol) to 3370 m (Yerdavletov, 2015) (Fig. 2.17).

The natural complex of light coniferous (Siberian larch) and dark coniferous (spruce and fir) taiga on slopes of the Kurchum range. In the forest there are many bushes: bird cherry, currant, hawthorn, dogrose. The mixed-grassy vegetation of mountain meadows is rich. Many plants have medicinal value (maral and golden roots). Wildlife of the reserve is rich and diverse. There are 39 species of mammals, about 200 species of birds. Among them are roe deer, brown bear, lynx, wolverine, squirrel-teleut, speakers, Altai ular and others (Yerdavletov, 2015).

The Rakhmanovs' spring are one of the most picturesque corners of the Southern Altai, they have long been well-known for their radioactive sources, which are close in composition to the waters of Tskhaltubo. It helped to discover these healing sources, as there is often the case. More than 200 years ago, a hunter Rakhman wounded a maral in the mountains. For two days he walked along the trail of an animal he had wounded until he found him standing in the clouds of steam. After shooting and killing maral, the hunter suddenly discovered that the wound from his

Fig. 2.17 Lake Markakol (*Resource* Available at https://Kazakhstan.travel/photobank)

first bullet on the leg of the beast was already tightened with a thin skin. And then the hunter tried to test the healing properties of a hot spring. Bathed three times a day, he recovered for a week from rheumatism, which he suffered for many years.

Resort named "Rakhmanovs' spring", which is very popular now for taking patients here. Warm nitrogen-radon baths and healing mountain air quickly restores people's health, contributes to the successful treatment of diseases on peripheral nervous system and other ailments.

Particular attention should be paid to the Kazakhstan Altai territory, which is part of the Altai-Sayan ecological region.

East Kazakhstan is well-known for its magnificent honey—Altai mountain honey for its aroma, taste and medicinal qualities and considered as one of the best in the world; here there are bears and marals (or noble deer). The diversity and beauty of the natural high-mountain landscapes, the richness of animal and plant world, the presence of turbulent rivers, noisy and rapid waterfalls, healing springs and other places of special prominence is the Katon-Karagai district.

Magnificence of the East Kazakhstan landscape is crowned by the highest peak of the Altai—two-headed Belukha mountain (4506 m), from which the Great and Small Berel glaciers descend. In the south Tarbagatay and Saur ranges rise.

The largest lake in the region is Zaysan, merges with the Bukhtyrma reservoir (Eastern Kazakhstan. About the regions of Kazakhstan. Tourism of the regions of Kazakhstan https://www.orexca.com/rus/Kazakhstan/regions/eastern_Kazakhstan.htm).

East Kazakhstan has all the favourable conditions for the ecological tourism development (Fig. 2.18).

Fig. 2.18 Belukha mountain (*Resource* Available at https://Kazakhstan.travel/photobank)

South Kazakhstan region (Turkestan region). South Kazakhstan region includes four administrative regions of the country: Kyzylorda, South Kazakhstan region, Zhambyl and Almaty. It is the second most occupied territory and the first in Kazakhstan by population.

The natural conditions of South Kazakhstan region in comparison with any other part of the republic are very diverse and have much in common with nature of neighbouring Central Asia. Here, as in Central Asia, there are vast arid plains and sufficiently moist mountains and foothills. And four fifths of the region territory is occupied by desert plains and only a small part—mountains and foothills. But in the economy of South Kazakhstan region, as in Central Asia, this small part of it plays a crucial role, since the highlands are covered with valuable subalpine and alpine meadows, where numerous rivers originate, and powerful deposits with fertile chestnut soils lie in the foothills. Here is the main part of crops, and the majority of population is concentrated and pass all transport lines. A narrow piedmont strip is the most developed and most important in the economical part of South Kazakhstan region.

Along the southern borders of Kazakhstan hundreds of kilometres stretch the mountain chains—the western spurs of the Tien Shan. They are interesting not only for the beauty of their peaks and gorges, but also for the fact that there are many architectural and archaeological monuments of Kazakhstan in these southern regions of Zhambyl and South Kazakhstan region.

In the South Kazakhstan region territory there are 5 reserves (Aksu-Zhabagly, Barsa-Kelmes, Almaty, Alakol and Karatau), 6 national parks (Altyn-Emel, Ile-Alatau, Sharyn, Sairam-Ugam, Kolsai lakes, Zhongar-Alatau), 17 the natural reserves and 4 nature sanctuaries (Sharyn ash-tree grove, Chinturgen spruce forests, Singing dune, Baum grove), 2 botanical gardens, 3 protected areas (Arys and Karatau, South Kazakhstan region, Zhusandaly) and 1 dendrology park—Issyk.

One of the most popular holiday spots for citizens and guests of region is lake Issyk.

Near it there is a magnificent gorge on the Turgen river. Here you can enjoy the mountain coolness, fresh air, walk to the waterfalls; see the well-known Sak burial mounds, dated to the 1000 c. BC. and relic groves that grew in permafrost conditions (Southern Kazakhstan. About the regions of Kazakhstan. Tourism of the regions of Kazakhstan https://www.orexca.com/rus/Kazakhstan/regions/southern_Kazakhstan.htm) (Fig. 2.19).

In the territory of Almaty region is the well-known Valley of Castles on the river Sharyn.

Among the most picturesque the natural landscapes can be considered the canyon of Sharyn river, a left tributary of the Ile river, flowing in the east of the region. The canyon is replete with towers, pillars, labyrinths of gorges. The "architecture" of it is so fantastic that the canyon is called the "Valley of Castles", the "Sphinx", the "Witches Gorge", etc. The Sharyn canyon resembles the well-known Grand Canyon of Colorado in the USA. Therefore, it is also called "mini Colorado". Here there are unique communities of plants and animals. These include the falcon-balaban, black stork, various snakes (Aliyeva 2002).

Fig. 2.19 Lake Issyk (photo taken by Artemyev A.)

On the territory of Almaty region there are objects of the greatest interest from the ecological tourism point. This is the Sharyn ash grove—the largest, unparalleled in the former Soviet Union, an array of relict tertiary plants. This grove is older than the canyon of the Sharyn river. The most valuable thing there is Sogdian ash—long-lived species. As well as the Shynturgen fir-trees, this is a relict part of dark coniferous northern taiga concerning glaciation era.

Kolsai lakes. Decoration of the Northern Tien Shan is Kolsai lakes located in the spurs of the Kungei Alatau. From the point of view of ecological tourism, the most interesting is Lake Kaindy, located near Kolsai Lakes (Fig. 2.20).

In the gorge, birches planted by human being rise to the sky, and perfectly adjacent to the coniferous landscape (Southern Kazakhstan. About the regions of Kazakhstan. Tourism of the regions of Kazakhstan https://www.orexca.com/rus/Kazakhstan/regions/southern_Kazakhstan.htm).

National park Altyn-emel, the largest reserve of Kazakhstan, is located on the western edges of Zhetysu Alatau range. The area of 460,000 ha is intended for the conservation of rare species of animals and plants. Here you can enjoy a unique view of the multicolored mountains of Aktau. On the park territory there are several historical monuments—Scythian (Sak) burial mounds of Bes-shatyr, dating from the 1st millennium BC and Tamgaly-tas with ancient rock images, the earliest of which are referred to 16–14 centuries BC (Southern Kazakhstan. About the regions of Kazakhstan. Tourism of the regions of Kazakhstan.

Fig. 2.20 Kaindy lake (*Resource* Available at https://Kazakhstan.travel/photobank)

https://www.orexca.com/rus/Kazakhstan/regions/southern_Kazakhstan.htm) (Fig. 2.21).

Aksu river canyon, located in the Aksu-Zhabagly nature reserve, is one of the most picturesque and deepest canyons in Central Asia (Fig. 2.22).

Descent into the canyon is quite complicated, and the steep banks of a river in many areas make it basically impassable. Rocks well warmed up by the sun and a mountain river providing the air humidity create a microclimate of the natural greenhouse in the canyon.

Glacier Tuyuksu, sparkling snow-white peaks, is the largest and most accessible glacier, located near the metropolis of Almaty. In this regard, it is very popular among ecotourists, both locals and guests of the southern capital.

It is located within the vast moraine field, where there are up to a dozen moraine lakes—foci of mudflows on glacial genesis. The basin of one of these lakes lies below the modern moraine of the Tuyuksu glacier. On July 15, 1973 there was a breakthrough of 180 thousand m^3 of meltwater, causing a catastrophic mudflow, detained by dam and silo on Medeu (Fig. 2.23).

After a steep climb through the moraine, in about 20 min, the "tongue" of Tuyuksu will appear, framed by a border of small stones. The setting here is such that tourists find themselves in the realm of rocks, snow and ice, and it seems that you are in the real Arctic. After passing another 500 m up the glacier, tourists finish the route in the middle part of "tongue", at an altitude of about 3600 m, where the zone of deep transverse cracks formed during movement of the glacier begins.

Fig. 2.21 Singing dune (*Resource* Available at https://Kazakhstan.travel/photobank)

Fig. 2.22 Canyon of Aksu river (own authors)

Fig. 2.23 Glacier Tuyuksu (photo taken by Artemyev A.)

Before going down, it is appropriate to remind the tourists that glaciers in the upper reaches of the Small Almaty were open to science about 100 years ago by the well-known local historian S.E. Dmitriev. In 1902, he made the first meteorological and geodesic observations on the main valley glacier Tuyuksu. Later more than 40 years it was studied by academician N.N. Palgov. Since 1956, Kazakhstan scientists have been conducting continuous systematic studies of the evolution, regime and mass balance of the Tuyuksu glacier according to international geophysical and hydrological projects. Thanks to these studies, now the Tuyuksu glacier is the most studied within the former USSR, and is among the five most studied glaciers of our planet and the most frequently visited by tourists (Aliyeva 2001).

Picturesque the natural landscapes of South Kazakhstan region are objects of ecotourism.

2.3 Protected the Natural Areas. National Parks, Reserves and Reservations

Specially protected the natural areas (SPNA) have the greatest potential for ecological tourism development.

Specially protected the natural areas include land and (or) sea areas purposely designated for the conservation and maintenance of biodiversity, the natural and associated cultural resources and have a special legal status.

The SPNA system is a combination of protected areas that are functionally and territorially complementary to each other, organized by considering the natural physiographic structure of the region and interactions with various forms of economic activity. Their objective is to ensure conditions for sustainable socio-economic development based on the preservation and improvement of environmental conditions, and preservation of biological diversity. They are subdivided into the following categories (Tourism of Kazakhstan 2015):

- state nature reserves are areas specially protected by law (territories or waters) completely and forever excluded from any economic activity, including visiting people. In reserves only scientific, security and control activities are allowed;
- the natural reserves—the word "reserve" in translation from the Latin language means "to save, preserve", this form of protected areas is most common abroad. In the form of protection, it has a reserved mode because the natural reserves make up the most valuable part of vast protected areas—haunting, national parks, reserves;
- national parks are areas of the territory (water area) from several thousand to several million hectares, including both fully protected areas and areas intended for aesthetic, recreational, scientific, cultural and educational purposes. In most countries of the world, national parks are the main form of landscape protection;
- state the natural reserves are the natural complexes (including water areas) on which for a number of years or permanently in certain seasons or certain types of animals, plants or part of the natural complex are protected year-round. The economical use of other resources is permitted in a form that does not damage

the protected facility or complex. Often, temporary hunting reserves are created to preserve and restore the number of any animal species;
- monuments of nature are separate irreplaceable the natural objects that have a scientific, historical and cultural-aesthetic significance. They can be caves, geysers, palaeontological objects, separate age-old trees;
- dendrology parks and botanical gardens are nature conservation institutions whose tasks include the creation of special plant collections, the preservation of diversity and enrichment of the plant world, as well as scientific, educational and outreach activities;
- therapeutic and recreational areas and resorts are territories (water areas) suitable for organizing the treatment and prevention of disease, as well as recreation of the population. These places have such the natural medicinal resources as mineral waters, curative mud, healing climate, beaches and other the natural conditions.

Present time in the Republic of Kazakhstan almost all the above categories of specially protected the natural territories are represented. The state protected the natural areas of the country include 10 state nature reserves, 12 state national the natural parks, 50 state the natural reserves, 26 monuments of nature; 5 botanical gardens; 5 state conservation areas, 5 state wildlife areas and 1 dendrology park (Table 2.1).

SPNA perform the following functions:

(1) direct and indirect use of environmental resources and biological diversity during economic activity;
(2) preservation and restoration of the environment and biological diversity as a result of special action program in the field of human relations with environment;
(3) informational support of the management system and society with knowledge of the existing reality and trends of its change;
(4) regulatory impact of the management system, coordinating the environment state and extent of its use and ensuring the sustainable development of society;
(5) formation of moral and ethical ideas that contribute to improving the efficiency of the natural resources use.
(6) In the Republic of Kazakhstan, there are 10 protected areas, 12 national parks and 5 nature reserves under active protection.

SPNA are the most important connection in the development of eco-tourism, because they have a number of prerequisites for this:

- located in places, which are the most picturesque, attractive, and interesting from an informative point of view;
- have already established a system of reception and maintenance on tourist groups, planned system of tourist routes, determined by the practice of organizing educational work;

Table 2.1 Distribution of SPNA in the natural areas of Kazakhstan (Constructed by author based on the book "Tourism of Kazakhstan")

SPNA zones	Forest-steppe and landscape of temperate zone	Steppe and landscape temperate zone	Semi-desert and landscape temperate zone	Desert and landscape temperate zone
1	2	3	4	5
Reserves (organization year)		1. Nauryzym (1937) 2. Markakol (1976) 3. West Altay (1992) 4. Korgalzhyn (1968)	1. Aksu-Zhabagly (1926)	1. Ustuyrt (1984) 2. Alakol (1998) 3. Almaty 1964) 4. Barsakelmes (1939) 5. Karatau (2004)
Nature reserves (organization year)	1. Mamlyut (1986) 2. Sogrov (1986) 3. Smirnov (1986) 4. Orlinogorsk (1986)	1. Mikhailovsk (1967) 2. Tounsor (1974) 3. Kuv (1986) 4. Belagash (1986) 5. Kyzyltau (1986) 6. Bulandin (1986) 7. Atbasar (1982) 8. East (2000) 9. Poima, Irtysh river (2001) 10. Low-Turgusun (1978) 11. Beldeutas (2008) 12. Zharsor-Urkash (2008) 13. South Altay (2012)	1. Zhaltyrkol (1967) 2. Budarin (1967) 3. Krysanov (1967) 4. Torgai (1967) 5. Karaagash (1986) 6. Bektauata (1986) 7. Turanga (1986) 8. Kyzylarai (1986) 9. Ulytay (1986) 10. Kulundzhun (1967) 11. Karatal sands (1986) 12. Tarbagatay (1967)	1. Kokagash (1986) 2. Novinsk (1967) 3. Karagie-Karakol (1986) 4. Aktau-Buzachi (1986) 5. Andasay (1966) 6. Karaoi (1979) 7. Pribalkhash (1957) 8. Kokan(1986) 9 Lepsy(1986) 10. Verkhnekoksu (1986) 11. Toty (1986) 12. Boralday (2001) 13. Zhambyl (1986) 14. Almaty (1953) 15. Karakonyz mountain area (1971) 16. Berikkara mountain area (1971) 17. Kargaly (1986) 18. Torangyl (1978)

(continued)

Table 2.1 (continued)

SPNA zones	Forest-steppe and landscape of temperate zone	Steppe and landscape temperate zone	Semi-desert and landscape temperate zone	Desert and landscape temperate zone
				19. Akdala (1986) 20. Zardarya (1986) 21. Timur (1986)
National parks (organization year)		1. Kokshetau (1996) 2. Borovoe (2000) 3. Katonkaragay (2001) 4. Karkaraly (1998) 5. Bayanaul (1985) 6. Buiratau (2011)	1. Sairam-Ugam (2001)	1. Ile-Alatau (1996) 2. Altyn-Emel (1996) 3. Sharyn (2004) 4. Kolsai lakes (2007) 5. Zhongar-Alatau (2010)
Wildlife reserves (organization year)		1. Ertis forests (2003) 2. Semey forests (2003) 3. Altyn-Dala (2012)	1. Yrghyz-Torgai (2007)	1. Akzhayik (2011)
Monuments of nature	1. Zhanazhol 2. Serebryanyi bor 3. Sosnovyi bor 4. Bald mountain of Orlinaya and springs 5. Lake island of Imantay 6. Bald mountain of Obozrenie 7. Rock deposit of Kotelok 8. Relict massifs 9. Steep bald mountain 10. Bald mountain of Dva brata 11. Waterfall with cave 12. Chapped bald mountain	1. Steep bald mountain of Shlem 2. Zelenyi mys 3. Man-made lake with relict plantations 4. Gummy bald mountains 5. Bald mountains of Strekach 6. Malinovyi mys 7. Bald mountains of cotton seed hulls 8. Bald mountains of Pozharnaya 9 Gusinyi perelet	1. Sinegorye—fir grove	1. Charyn ash-tree forest cottage 2. Chinturgen fir-trees 3. Singing Dune 4. Baum grove
Protected areas			1. Arys and Karatau (2001)	1. North part of Caspian Sea

(continued)

Table 2.1 (continued)

SPNA zones	Forest-steppe and landscape of temperate zone	Steppe and landscape temperate zone	Semi-desert and landscape temperate zone	Desert and landscape temperate zone
			2. South Kazakhstan region 3. Kenderly–Kayasan (2001) 4. Zhusandala	
Botanical gardens			1. Zheskazgan 2. Mangyshlak	1. Ile 2. Main Almaty 3. Altay
Dendrology parks				1. Issyk (2016)
Total SPNA zones	16	35	22	41

- have a certain infrastructure, transport, personnel and communication systems;
- local population has already developed an established attitude towards a specific the natural reserve and environmental restrictions on economic activity existing on its territory (Table 2.2).

Table 2.2 State nature reserves of Kazakhstan (constructed by author based on the book "Tourism of Kazakhstan")

№	Name of the reserve	Organization year	Total area, ha	Administrative region	Main landscapes
	Aksu-Zhabagly	1926	131,934	South Kazakhstan region	Mountain, forest
	Almaty	1964	71,700	Almaty	Mountain, forest
	Barsakelmes	1939	172,070	Kyzylorda	Deserts
	West–Altay	1992	56,078	West-Kazakhstan	Mountain, forest
	Korgalzhyn	1968	258,963	Akmola	Wetland
	Markakol	1976	75,048	East Kazakhstan	Mountain, forest, lake
	Nauryzym	1934	191,381	Kostanay	Lake, steppe, forest
	Ustyurt	1984	223,342	Mangistau	Deserts
	Alakol	1998	197,129	Almaty	Wetlands
	Karatau	2004	34,300	South Kazakhstan region	Deserts

Fig. 2.24 The Almaty reserve (photo taken by Artemyev A.)

Reserves play one of the decisive roles in the preservation of rare and valuable plants, wild animals, birds, fish, as well as places of fish breeding grounds, wintering, nesting waterfowl, etc. (Aliyeva, 2002). Thus, in the **Almaty reserve,** foothill steppes, spruce forests and alpine landscapes of the Trans-Ile Alatau were taken under protection. Almaty reserve is located in Talgar district of Almaty region (Fig. 2.24).

Its main value is the dark coniferous forests of Tien Shan spruce. The trees of this breed are durable and high—the Tien Shan spruce lives for almost 500 years and reaches 50 m in height. It can tolerate dry air, heat, cold, lack of moisture in the soil; such unique properties of spruce wood have long allowed using it as a building material. The acoustic properties of fir wood, which is used as a valuable raw material for the manufacture of musical instruments, are widely known. At an altitude of 2500–3000 m thickets of creeping juniper are common, which have the ability to trap avalanches. The Almaty reserve is a habitat for the snow leopard, the brown Tien Shan bear, and they are under protection of the World Conservation Union. Here you can also find nests of the world's rarest bird—sicklebill, the red-headed shakhin falcon, and also the blue bird nests, bearded vulture and big lentils (Aliyeva, 2002).

Fig. 2.25 The Markakol reserve (*Resource* Available at https://Kazakhstan.travel/photobank)

In the **Markakol reserve,** South Altai middle-mountain spruce, fir and deciduous forests with fragments of mountain steppes are taken under protection. The reserve is located in Markakol district of East Kazakhstan region, in the south-east of Southern Altai (Fig. 2.25).

The reserve territory is represented mainly by Markakol lake, located at an altitude of 1485 m. The colour scheme of lake amazes with its amazing diversity and beauty. Most often, the lake is blue or light blue, however, with changing weather takes shades of gray, black, silver. Lake is inhabited by endemic fish species, such as goldilocks, which has the local name "uskuch", grayling, char, gudgeon. In addition to the indigenous population—the Kazakh, in the nineteenth century, the lake was visited by peasants from Bukhtyrma village, who annually extracted and exported up to 50 thousand pounds of fish. Massive fish catch led to the impoverishment of fish stocks in lake (Aliyeva 2002).

In addition to the lake, landscapes of larch, fir and spruce taiga typical for the Altai Mountains are of great value. Reserve has a lot of valuable medicinal and industrial plants typical for the region. Here there are steppe peonies, narrow-leaved willow-tea, white-coloured geranium, pink radiola (golden root), Altai rhubarb, which is endemic, and Siberian kandyk. The fauna of reserve is characterized by the presence of red deer, roe deer, elk, mountain goat, American mink, and sable.

There is a great interest of the **Nauryzym reserve**, located in the Kostanay region. Its area is 87.7 thousand ha. It was founded in 1934 for study and protection of waterfowl, as well as for the restoration and protection of steppe forests, such as

Fig. 2.26 The Nauryzym reserve (*Resource* Available at https://Kazakhstan.travel/photobank)

the relict of Nauryzym steppe forest and the strip forest Tersek (Aliyeva 2019) (Fig. 2.26).

Here is the southernmost boundary of pine growth and unique Nauryzym forest. Above the reserve runs the great Torgai bird migration route. In Nauryzym reserve, besides relict pine forest among dry fescue-feather grass steppes, salt and freshwater lakes are protected. The reserve also owns wetlands of international importance. Landmarks of the lakes are large colonies of curly and pink pelicans, as well as nesting places for white cranes, gulls, gray geese and ducks (Aliyeva 2002).

The **Korgalzhyn reserve** is located in the Tengiz-Korgalzhyn depression, which is occupied by the lakes Tengiz and Korgalzhyn in the Akmola region. Here, as in the Nauryzym Reserve, there are wetlands of international importance. The reserve was founded in 1968 to protect the nesting sites of waterfowl.

The Korgalzhyn reserve is the only reserve in Kazakhstan included in the Convention on the Migratory Birds Conservation, whose lakes are recognized as wetlands of international importance. The nest of pink flamingos in Eurasia is guarded here. In spring, with the arrival of birds, bird markets are collected in the reserve. Particularly noteworthy are such species of birds as the black stork, the belladonna crane, curly pelican, and the swan-loaf. Korgalzhyn reserve is deservedly called the land of fearless birds. In spring and autumn, more than three million of them rest here (Fig. 2.27).

Fig. 2.27 The Korgalzhyn reserve (*Resource* Available at https://Kazakhstan.travel/photobank)

Reserve is located in the steppe zone. Therefore, there are such steppe plants as red and purple tulips, blue irises, yellow buttercups, purple saffron, colourful carpet covering the steppe expanses of reserve in spring. Unique plant of the Korgalzhyn reserve is a plant with mysterious name—dream-grass—the charming beauty of dark purple bell-shaped flower. According to beliefs, this spring flower can cast dreams. At the same time, numerous herds of wandering saigas pass through the western part of the reserve (Aliyeva 2002).

The *Ustyurt reserve*—established in 1984 in territory of the Karagie district in the Mangystau region to preserve the entire the natural complex of south-western Kazakhstan. Its area is 223.3 thousand ha. It is located on slopes of the Ustyurt plateau and covers the horseshoe Kendirly. Ustyurt is a vast table plateau with a series of elevations, gentle ridges and depressions located between the Aral and Caspian Seas. Geologically, the topography of plateau is unique. Suddenly, against the background of plain with a wavy surface, an almost vertical break appears. The western chink encompasses in form of a giant horseshoe on the Mangyshlak plateau from the east. This is a unique the natural phenomenon of erosion character with peculiar relief forms that have no analogues in the world, in the form of vertical 200-m walls, remnants of table-like and conical shapes. Chinks are represented by outcrops of limestone and chalk, have the most diverse shape and colour range, from pale pink and blue to dazzling white. Here, aeolian relief forms, extensive dry depressions, dry beds of modern and ancient temporary streams are widespread (Aliyeva 2019) (Fig. 2.28).

Fig. 2.28 The Ustyurt reserve (*Resource* Available at https://Kazakhstan.travel/photobank)

Reserve is protected by typical desert landscapes of the Turan lowland and Ustyurt plateau. Flora and fauna of the reserve are typical for northern deserts; all the listed species of animals and birds, mainly listed in the Red Book of Kazakhstan, live here. In winter, saigas migrate to these places. Jacks, black-sprouted hazel grouses, saker falcons are nestling here. The flora is poor, which is represented by ephemera and hodgepodge (biyurgun, wormwood, tamarisk, saxaul, etc.).

Caravan routes connecting Khiva with lower reaches of the Emba and Volga passed through the reserve. Therefore, a landmark of this area are the archaeological monuments, represented by ancient settlements, ruins of fortresses, ancient mausoleums, some of which have not yet been investigated. All of them are objects of ecological tourism.

The state the natural *West Altai reserve* was established in 1992 on 86,122 ha of north-east on the East Kazakhstan region territory of two administrative districts: Ridder and Zyryan.

In the West Altai Reserve, the middle mountains with dark coniferous taiga, alpine and tundra highlands of the West Altai are protected. A unique formation is granite remnants of the Liney ridge, representing the original geological monuments of past ages, formed by time and precipitation. Characteristic forms of relief are kars—remnants of upper-quaternary glaciation, reaching a height of 100–150 m. (Aliyeva 2019) (Fig. 2.29).

Fig. 2.29 The West Altai reserve (*Resource* Available at https://Kazakhstan.travel/photobank)

Forests of the West Altai have a great importance for the ecological tourism development. Chernevaya taiga, one of the leading local formations, differs from other phytochromes by a number of features—predominance of fir in the first tier, aspen and birch in the second tier, and cherry-tree, viburnum, and mountain ash as additional. Cedar-fir forests are transitional formation between blackened and dark coniferous taiga. Dark coniferous taiga is distributed at an altitude of 1500 m and is represented by species such as spruce, cedar, fir. In the reserve there are several formations of tundra, formed mainly by pine species, such as the Siberian kandyk, snowy long-legged, and the Alpine moss.

Due to the fact that the West Altai reserve was formed quite recently, its fauna, in comparison with fauna of other reserves in Kazakhstan, is the least studied. The most complete are information about such "Red Book" species of birds, as the black stork, saker falcon, golden eagle, merlin, owl, and black grouse. Among the mammals, there are American mink, roe deer, sable, brown bear, and maral (Aliyeva 2002).

Barsa-kelmes reserve is located in the Aral region of the Balkhash-Aral climate zone. In reserve, a deserted island in the Aral Sea is taken under protection with wormwood, sunflower vegetation and reed beds. Reserve was created to restore the koulans. Barsa-kelmes, formed in 1939, is one of the oldest reserves in Kazakhstan. The name "Barsa-kelmes" in translation means "If you go, you will not return." (Aliyeva 2002) (Fig. 2.30).

The reserve is an experiment site on acclimatization of the Turkmen koulan—an animal for which any hunting is prohibited. Koulan is a wild donkey, and although in its appearance it looks more like a horse, but somewhat smaller than it and with a short tail. Fauna of Barsa-kelmes, in addition to the koulan, is represented by hare,

Fig. 2.30 The Barsa-kelmes reserve (*Resource* Available at https://Kazakhstan.travel/photobank)

pheasant, gray partridge, gopher-sandstone, steppe turtles, and lizards. Flora of the the island is represented by perennial shrubs and shrubs, which are typical for the desert, xerophytes, ephemeras. Desert forests of black saxaul are a decoration of island and an ecological shelter for many animals. They are found in the main landscapes of the skeleton and are quite diverse. The most valuable protected animals of the reserve are ungulates: saigas, gazelles and koulans, which adapted to the harsh desert conditions of an island and became the main components of its biocoenoses.

One of the most interesting the natural sites in South Kazakhstan region is the oldest reserve of the republic—*Aksu-Zhabagly,* created to protect the highland landscapes of Talas Alatau, juniper light forests and floodplain forests. It was created in 1926, its area has been increasing all the time, and now it is 132.6 thousand ha (Aliyeva 2019) (Fig. 2.31).

It is located on the Talas Alatau, Ugam and Maidantal ridges, in the west part of the Tien Shan. Reserve name is formed from the name of its two largest rivers—Aksu and Zhabagly, in the interfluve of which it was created. Aksu-Zhabagly is a small part of huge mountainous country—the West Tien Shan, which has a peculiar and unique nature. There is a sharp change of relief—from the mountains and gorges—to the steppes and semi-deserts. Amazing animals and plants inhabit these mountains.

The most valuable mammals are ungulates: argali, mountain goat, roe deer, deer. Bearded vulture has a great interest as one of the largest eagles, reaching up to 3 m in wingspan; here you can also find nesting sites of keklik and ular—mountain turkey, falcon, baloban and griffon vulture, etc. Flora of the reserve is diverse,

Fig. 2.31 The Aksu-Zhabagly reserve (*Resource* Available at https://Kazakhstan.travel/photobank)

among which there are a tree-like juniper up to 10–12 m high, persian rowan, walnut tree, almond, etc. (Aliyeva 2002).

State nature *reserve Alakol* is organized in the Tentek river delta, where many unique and very valuable biocenoses coexist. Thus, rare populations of relic seagulls and other colonial birds that have found their refuge on the islands of lake Alakol are subject to special protection (Fig. 2.32).

Reserve itself is located in Alakol and Urdzhar districts of Almaty and East Kazakhstan regions. It is named after the Alakol lake located on its territory (Aliyeva 2019).

Delta wetland landscapes, coasts of lakes and islands on intermountain desert depression are located in the centre of Eurasia and protected in the Alakol reserve.

State nature reserve Karatau is the youngest among the republic reserves and it was established in 2004.

The *Karatau reserve* is located in the central part within the Karatau range, which is an offshoot of northwestern arches in Tien Shan on the territory of the South Kazakhstan region. It borders with deserts of Moyunkum, Kyzylkum, Betpak-dala. The climate is continental and arid. Syrdarya Karatau has the peculiarity of its vegetation cover. In general, reserve flora of higher plants, according to preliminary data, has more than 400 species. At least 25% of the species total number belongs to the category of rare. Only Karatau endems in the modern territory of reserve are 53 species, another 10 are endemic to the Karatau and West

Fig. 2.32 State nature reserve Alakol (*Resource* Available at https://Kazakhstan.travel/photobank)

Tien Shan. In the Red Book (considering the prepared 2nd edition) 42 species are listed. According to the abundance of endemic and rare plant species, the Karatau ranks as first in the republic (Fig. 2.33).

The fauna of the protected area, as well as the Karatau range in whole, has undergone significant changes over the past century. During this period, the Tien Shan bear and snow leopard were destroyed here, and many commercial species (wild boar, roe deer, argali, tolai hare, etc.) became extremely few and far between. Only in Karatau, there is a rare and endangered population of Karatau argali. On the territory of Karatau reserve there are 3 species of mammals: argali, Indian porcupine and stone marten. Avifauna contains 118 species. There are 12 species of birds are endangered in the Red Book of Kazakhstan (1996), which reduced and considered as rare (Tourism of Kazakhstan 2015).

In created reserves, first, as reserves of the natural complexes, extensive scientific and research work is carried out. This gives rise to the use of protected areas for scientific ecological tourism. The natural landscapes of reserves are objects for obtaining ecological knowledge, ecological training, which is possible within the framework of scientific ecotourism. Scientific ecotourism is organized mainly for scientists, environmental specialists, environmental students, geographers, tourists.

National nature parks are a functional network in which environmental objectives are combined with interests of environmental education, tourism and recreation of the population. In general, national parks are geared towards meeting recreational needs.

Fig. 2.33 State nature reserve Karatau (*Resource* Available at https://Kazakhstan.travel/photobank)

Objects interesting for ecotourists in national parks are the natural landscapes, plant and animal resources, gene pool of rare and endangered organisms, unique the natural complexes.

In terms of their functional purpose, national parks meet the objectives of ecological tourism, which primarily include:

- meeting people's needs from direct communication with nature;
- recreation in nature with aesthetic pleasure from the natural and cultural values.

To ensure the functioning of ecological tourism in the territory of parks, while strictly observing the environmental regime, the park administration must clearly organize ecotourism routes, allocating special ecotourist zones, trails for the inspection of picturesque landscapes, nature monuments and other remarkable objects.

There are currently 12 national parks in Kazakhstan (Table 2.3).

The *Ile-Alatau national park* is located in the north part of the Zailiysky Alatau ridge, and characterized by the most attractive the natural complexes. The wonderful nature of mountains, especially in Small Almaty, is so concentrated that in one day, ascending from the city of Almaty to the Tuyuksu glaciers, person can go through almost all-the natural areas (Fig. 2.34).

Table 2.3 State national the natural parks (constructed by author based on the book "Tourism of Kazakhstan")

№	Name of the state national the natural parks (SNNP)	Organization year	Total area, ha	Administrative region	Nature zones
1	Bayanaul	1985	50,688	Pavlodar	Steppe
2	Kokshetau	1996	182,076	Akmola	Steppe
3	Katon-Karagay	2001	643,477	East Kazakhstan	Steppe
4	Karkaraly	1998	90,323	Karagandy	Steppe
5	Borovoe	2000	83,511	Akmola	
6	Sairam-Ugam	2006	149,053	South Kazakhstan region	Semi-desert
7	Ile-Alatau	1996	199,703	Almaty	Desert
8	Altyn-Emel	1996	161,153	Almaty	Desert
9	Sharyn	2004	93,150	Almaty	Desert
10	Kolsai lakes	2007	161,045	Almaty	
11	Zhongar-Alatau	2010	356,022	Almaty	
12	Buiratau	2011	88,968	Akmola, Karagandy	Steppe

Fig. 2.34 "Ile-Alatau" national park (photo taken by Artemyev A.)

Undoubtedly, the natural complexes of the Zailiysky Alatau mountains are of great ecological, scientific, and aesthetic value. Aspen groves and wild fruit forests are subject to protection here. Spruce forests are the natural phenomenon of coniferous trees deep movement in the arid zone. It is here that the forests of Schrenk spruce grow, as well as the preserved wild fruit trees, which are based on the Sivers' apple tree, and according to scientists, is the ancestor of cultivated apple varieties. There are about 200 species of vertebrates in the park, which makes up 23% of the Kazakhstan fauna. However, with such a diversity of wildlife, most of them are the objects of the World Conservation Union. These include: argali, or Tien-Shan mountain sheep, snow leopard, Central Asian otter (in whole Kazakhstan lives only in the upper Ile river), Tien-Shan brown bear, porcupine. Ile-Alatau park is the only place where the rarest bird nests that attracts attention of ornithologists from all over the world—the sicklebill.

On the territory of park there are archaeological monuments of different epochs that are fascinating for ecotourists. Among them: 2 medieval settlements, 10 early iron kurgan burial mounds, 4 settlements of the early iron era, and 3 sites of rock art. The "royal" burial mounds of eastern Scythians, as well as the remnants of medieval cities and settlements located along the Great Silk Road linking cultures of East and West, and considered as an asset of national park.

Position of the park, its functional features, diversity and picturesqueness of landscapes, as well as the strict environmental protection regime provides for organization of ecological tourism (Aliyeva 2002).

Original and unique from the point of view of ecotourism is the *national park Altyn-emel* (Fig. 2.35).

There are 60 species are rare and endemic in the park. The park is one of the few places where there is a large concentration of wild animals per unit area, and therefore, a free view of wild animals from a car window is possible. So, there are about 4000 individuals of the desert antelope—gazelle, about 3000 mountains Siberian goats tau-teke, more than 300 koulans, descendants of koulans, imported from Barsakelmes island in the 80 s.

On the territory of park there are unique the natural formations and monuments of ancient culture, which are objects of ecotourism: "Singing dune", reaching a length of three kilometres and over 80 m in height, Aktau mountains, forming a "lunar" landscape, as well as five mounds of Beshatyr, representing historical burial of Sak kings, whose age is approximately 2500 years. In the gorges you can find cave paintings of ancient people.

Particular interest is the unique singing dune called Akkum-Kalkan, which makes a sound that resembles organ music. The sandy ridge contains quartz and calcite. Its upper layer, which is free-flowing and dry, while creating sand under the action of wind, creates such peculiar sounds (Aliyeva 2002).

Territory of *Bayanaul national park* is located within the Kazakh small mountain area, which, having emerged as a large mountainous country still in the upper paleozoic, has experienced a long history of continental destruction and therefore currently has relatively small heights (from 400 to 1027 m above sea level). The highest point of Bayanaul (1027 m) is mount Akbet.

Fig. 2.35 The national park "Altyn-Emel" (*Resource* Available at https://Kazakhstan.travel/photobank)

The mountain massif is considered as complex mining range, consisting of separate groups on various formed rocky hills, is the most attractive the natural oasis in all of North Kazakhstan. Here are pine forests with a mixture of birch and lush meadows. In the intermontane basins lie the freshwater lakes Zhasybay and Sabyndykol, from the waters of which rise bizarre rocks, where people have long since composed legends and songs. The Bayanaul mountain forest oasis has an area of 500 km. In addition to coniferous forests, there are also small areas of larch forests in floodplains of the Ertis, Yesil and Tobyl river valleys.

In the national park "Bayanaul" the Bayanaul mountains composed of granites with various forms of weathering are fascinating for tourists. Many rock formations have their names: "Golub'", "Baba-yaga", "Stone feather-beds". Picturesque landscapes are created by the beautiful lakes Zhasybay, Toraigyr, located among the mountains. Climatic conditions of Bayanaul are favourable for human health: there is clean air, many springs, lakes, beautiful valleys and gorges with thickets of raspberries, black currants, and blackberries (Aliyeva 2002).

Significant interest for ecotourists has Kokshetau mountains. This complex mining combination consists of separate groups of various formed rocky hills. "Earth, as ringing horseshoe"—this is how local akyns call the Kokshetau (Sinegorye) mountain range. It is not so high. The height of individual hills does not reach 200–300 m. The highest peak of this horseshoe-shaped ridge is Sinyukha (1047 m above sea level). From it, you can see the majestic panorama of the fairy-tale lake Borovoe surrounded by green pine forests and amazing granite rocks. Okzhetpes rock composed of large granite slabs is especially beautiful. The most amazing in its bizarre form is the rock "Zhumbaktas" ("Sphinx"), rising from the emerald waters of the lake Borovoe, it fascinates every tourist. And in many other rocks turned by winds and rains, the art of nature is seen. This is the well-known resort Borovoe, in Kazakh called Burabay. This is a truly blessed land of silence, crystal clear air, blue mountains and emerald lakes (Fig. 2.36).

The state the natural park Borovoe was created with the aim to protect the lake-mountain-forest landscapes, streamline tourism and organize recreation in 2000. The park is located in the territory of the Shchuchinsk district in Akmola region. This recreational Zhem of Kazakhstan is located in a fantastic area filled with mystical wonders. Local guides call it "Switzerland of Kazakhstan".

Borovoe climate is dry and healthy. There are 14 lakes in the park, with a water surface area of about 1 sq. km each, and a large number of smaller lakes. Large lakes are Borovoe, Shchuchye, Kotyrkol, Big and Small Chebachye, Karasye, Svetloye and others. Most of the lakes are fresh, some are salty. Searching for moisture, they picturesquely fled along the mountain slopes, stuck in the cliff crevices, in a wild pile of granite stones. There are birches, yellow acacia, cherry, and elm grow beautifully here. In forest glades, an abundance of wild currants, cranberries, cluster berries. Local forests are rich in medicinal herbs and mushrooms (Fig. 2.37).

Fig. 2.36 The national park "Bayanaul" (*Resource* Available at https://Kazakhstan.travel/photobank)

Fig. 2.37 The state national the natural park "Borovoe" (*Resource* Available at https://Kazakhstan.travel/photobank)

Exotic rocks such as "Okzhetpes", "Sphinx", "Camel", which are the most important monuments of the inanimate nature of this region, are very intriguing for ecological tourism. Mountainous riches of the Borovoe environs open unexpected secrets for ecotourists. Among them, the picturesque Okzhetpes rock (translated as "inaccessible to an arrow"), the top of which resembles a baby elephant, and the Zhumbaktas rock rising directly from the bay water (translated as "riddle") resembles a mysterious sphinx, interesting because when viewed from different points, then you can see the face of a girl with flowing hair, then—women, and, finally, old women.

The name "Borovoe" comes from the word "Bura", which means "camel". People kept the poetic parable of camel, which lived in the Kokshetau vicinity. It was peacefully grazing on the mountain slopes, when, sensing the enemy hordes approach, it flew up to the mountain top and with a powerful roar warned of impending danger. One of the Kokshetau mountain ridge rocks with its outlines resembles a double-humped camel from this beautiful legend.

The state national the natural park Kokshetau was organized by decision of the Republic of Kazakhstan government in 1996 to preserve and restore unique mountain-forest systems, historical, archaeology and national culture monuments of North Kazakhstan. Its main territory includes mountain forests and lake areas of

Zerenda, Shalkar, Imantau and Aiyrtau the natural zones. The administrative centre of the park is located in Kokshetau city.

Territory where the park is located is surprising for its beauty. Combination of picturesque glades, bordered with birch and pine pegs, high hills with rocky outcrops of trees and shrubs, with numerous lakes, marshes create the unique charm of these places. Each lake, glade, hill, spring has its name. Among people there are many legends, beliefs about the sacred birch, hills and other the natural monuments.

Large granite massif causes the development of local microclimate.

Medical resource fund is wealthy, including a complex of climatic factors that, with a source of mineral water and deposits of therapeutic mud, have a favourable effect on the physiological state of a person and contribute to the treatment of many diseases.

The state national the natural parks "Kokshetau", "Bayanaul" and "Borovoe" are a convenient and promising territorial form of nature conservation, recreation and tourism for North Kazakhstan (Fig. 2.38).

Among the steppes of Central Kazakhstan, there are small arrays of the strongest rocks: Kuu, Kent, Karkaraly, Kyzylrai. The *Karkaraly state national the natural park* was established on December 1, 1998 in order to protect mountainous forest landscapes and unique objects of nature, develop tourism and improve recreation. It is located on the territory of the Karkaraly district in Karaganda region. Ecotourists here are attracted by the low mountains of Karkaraly and Kent, located in east part of the Kazakh hillocky area. The highest point of the Karkaraly mountains is 1403 m, the Kent mountain range is 1469 m. In distant past, more powerful ridges

Fig. 2.38 The state national the natural parks "Kokshetau" (*Resource* Available at https://Kazakhstan.travel/photobank)

rose here, which over time collapsed and divided into many small ridges and hills. Their steep slopes are heavily dissected by erosion, and rocky ridges alternate with deep ravines and valleys.

Granite slabs exposed in many places seem to be superimposed on each other in the form of stone "mattresses", and some rocky outcroppings have the most bizarre outlines. A special feature of the flora is the large presence of relict plants belonging to the ice age; rare and endangered plant species—66 species. In the park located a monument of medieval history—the Kyzyl-Kensh palace, which, according to scientists, it was a Buddhist sanctuary (Tourism of Kazakhstan 2015).

Small lakes of Pashennoe, Bassein, Shaitankol are charming with their beauty (Fig. 2.39).

Mountains are surrounded by fabulous beauty, purity of the mountain air, filled with pine forest aroma. Peaks of the Karkaraly mountains do not go beyond the clouds, their maximum height is 1400 m above sea level, but they are well-known for their unique landscape—birch trees growing on rocks and cliffs, talkative transparent streams, cutting pine forests; clean spring sources, lakes. There are a lot of mushrooms and berries. Strawberries ripen on open sunny glades, currants and raspberries grow in shady places.

Among the remarkable landscapes are the mountains Ermentau, birch-aspen woodlands in the Mugalzhar mountains. There are many picturesque landscapes, lakes and river valleys in the steppes of Kazakhstan. Aksuat and Sarymain lakes

Fig. 2.39 The Karkaraly state national the natural park (*Resource* Available at https://Kazakhstan.travel/photobank)

serve as a resting place for migratory waterfowl. On the lake Seletyteniz rare birds nest.

The *Katon-Karagai national park* is the largest specially protected the natural area in Kazakhstan, its area is 643,477 ha.

Park includes the slopes of the highest peak of Altai–Belukha mountain (4506 m above sea level), Southern Altai (the highest point of Dzhagyrtau city—3871 m), Tarbagatai (Crystal city—3094 m) and Sarymsakty (Berkutaul city—3373 m). The high recreational potential of territory with a large number of objects on the natural, historical and cultural significance, as well as the transboundary location of other specially protected the natural territories in Russia, Mongolia and China, led to the creation of specially protected the natural area (Tourism of Kazakhstan 2015) (Fig. 2.40).

There are species like peony chimic, Altai rhubarb, radiola pink, sundew round-leaved, altai siberian, bearberry, etc. Many plants have various beneficial properties. Medicinal herbs are widespread: badan leaf, tea penny, hopper's Hypericum, yarrow, aconites, and many others. In some years there are incredibly large harvests of mushrooms, "pine nuts" and various berries: Altai honeysuckle, strawberries, red and black currants, cowberries bilberries, as well as the rarest of small-fruited cranberries in Kazakhstan, which are included in the Red Book.

Fauna of the park is very diverse. At least 15 species are listed in the Red Book lists: from the fish—taimen; from the nesting birds—black stork, hook-nosed turpan, osprey, burial ground, golden eagle, saker falcon, belladonna crane, owl; from

Fig. 2.40 The Katon-Karagai (*Resource* Available at https://Kazakhstan.travel/photobank)

the mammals—snow leopard, stone marten, Altai mountain sheep (argali). A unique industry was born here as maral breeding. From the horns of deer receive valuable pharmacological raw materials.

On the territory of the Katon-Karagay park there are Berel burial mounds, which are of great interest for ecotourists. They found unique finds of nomadic elite in the Scythian-Sak period (4th century BC), preserved due to the permafrost phenomenon—mummified remains of people and horses.

Large number of objects and ornaments were found here, felt fabric and dressed wool, the remains of horses with harness, and burial room with a large number of different objects. Finds made during excavations has global importance.

The territory of *Sharyn national park* includes a canyon-shaped valley and fan-shaped river delta, adjacent desert areas of the Syughatinsk valley and the low mountain range Ulken-Boguty, as well as fragments of the steppes in upper part of the valley and on slopes of the Toraigyr range (Fig. 2.41).

Sharyn canyon "Valley of Castles" is a geomorphological object, clearly reflecting the processes of relief formation and having special value for ecotourism. Numerous beams and ravines, wriggling, intersecting and again separating, form a dense disordered network, the so-called badland. The badlands and other areas of parks include many forms of relief—low mountains, small hills, plains, sandy massifs, slopes to the canyon and canyon itself with numerous terraces, shelves, caves, grottoes, odd single rocks, etc. Landscapes as a whole are peculiar—mountain, foothill and intermountain-hollow, as well as their individual unique components. Among unique are the ash grove in the Sarytogai tract, the red and

Fig. 2.41 The state national the natural park "Sharyn" (*Resource* Available at https://Kazakhstan.travel/photobank)

yellow canyons of Sharyn, the Valley of Castles, peculiar badlands formed by deposits of the Ile suite, stony gadam of the lower depression part, etc.

Ash Grove, located in tract Sartogai, has the status of "nature monument". Ash is a moisture-loving tree of amazing beauty. It grows about 300 years old, reaches a height of 25 m, and the diameter of its trunk exceeds one and a half meters.

In the Sharyn canyon, thanks to a rare combination of geological factors, interesting remains of prehistoric flora and fauna representatives have been preserved. In the cliffs of Sharyn there were found rich accumulations of fossilized bones (mastodon, elephants, rhinos, Stenon's horse, etc.), as well as original floristic pollen complexes. The Red Book includes an eagle owl, black stork, golden eagle, burial ground, eagle-dwarf, snake eagle, saker falcon, etc.

Sairam-Ugam national park was opened in 2006. The park was created by merger of the Ugam, Tolebi and Tulkubass state institutions for the protection of forests and wildlife.

There is a bizarrely shaped rock of Kyrykkyz—Forty Girls, Greyga's fabulous beauty tulip and much more. On the territory of park, magnificent closed juniper forests have been preserved in the valleys of Sairamsu and Ugam rivers.

Tien Shan is the northernmost of the great mountain systems of Inner Asia. Through the Dzhungarian range and Tarbagatay, it is associated with Altai and other mountains of Southern Siberia and Mongolia, through the Pamir, Kunlun and Karakorum—with Tibet and the Himalayas, and through the Pamir and Hindu Kush —with the Kopetdag and Iranian highlands. All these connections enrich the flora and fauna of the Tien Shan, make it at the same time typical of the Central Asian mountains and quite unique in its composition (Fig. 2.42).

This uniqueness also has global significance, since a large number of endemic plant and animal species of the Western Tien Shan are concentrated here. It can be emphasized that almost all Red Book species are rare.

The *Kolsai lakes state national the natural park* was established by the Republic of Kazakhstan government decree in 2007. Total area is 161045 ha. 72% of the territory is protected, 13% of the area is a zone of recreation and tourism (Fig. 2.43).

On the territory of park there are unique landscapes, with a cascade of lakes, rich in flora and fauna. Three unique mountain tectonic lakes are the main attraction of national park. They are called the pearl of the Northern Tien Shan, located in Almaty region. Lakes are located 120 km south-east of Almaty (Tourism of Kazakhstan 2015).

Kolsai lakes are a unique object of ecotourism. The lakes are surrounded by cliffs, coniferous forest from the Tien Shan spruce, forest-diversified tree species and mountain alpine meadows. Almost all types of flora species grow here. Rainbow trout lives in the waters. The big difference in temperatures allows ecotourists to swim and engage in fishing, horse riding, hiking in the summer, and skiing and skating on the lake ice in winter.

There is a fantastic lake Kaindy in the territory of park, located at an altitude of 1867 m above sea level surrounded by mountain peaks. Here is the abode of the most beautiful birds and animals, animal and plant world is very diverse.

Fig. 2.42 Sairam-Ugam national park (*Resource* Available at https://Kazakhstan.travel/photobank)

Fig. 2.43 The Kolsai lakes state national the natural park (authors)

Among them is edelweiss. Edelweiss is a mountain flower, very rare species, which is legendary, grows at the very top.

Animal world is rich and diverse, too. There are more than 200 species of vertebrate animals. As well as four species of fish, two species of amphibians, 197 species of birds and 29 species of mammals (Tourism of Kazakhstan 2015).

Monuments of history and culture are particularly interesting for ecological tourism. These include petroglyphs "Tanbaly Tas", the site of ancient settlement Baryskhan. In the gorge Tanbaly Tas there are ancient images (petroglyphs) of hunters with bows who hunted mountain goats. The settlement of Baryskhan is the location of ancient Baryskhan city, where Muhammad Kashgari was born.

The *Zhetysu-Alatau state national the natural park* was created in 2010 by the Republic of Kazakhstan government decree.

The territory of Zhetysu-Alatau park is a beautiful corner of land, which is attractive for domestic and foreign ecotourists. Unique the natural conditions of territory—the favourable climate and significant aesthetic attractiveness of mountain landscape are supplemented by the wealth and diversity of plant world, especially picturesque coniferous forests and significant number of red-knit species concerning animals living in region. The park's flora is represented by 2168 plant species (of which 76 are endemics inherent only in the territory of Zhetysu-Alatau, including rare and endangered types, listed in the Red Book of Kazakhstan, such as: Sivers apple-tree, pale-flowered grouse, Dzhungar gentian, thyroid veronica, etc. (Tourism of Kazakhstan 2015) (Fig. 2.44).

Fig. 2.44 The Zhetysu-Alatau state national the natural park (*Resource* Available at https://Kazakhstan.travel/photobank)

Zhetysu Alatau is one of the most saturated regions of Kazakhstan in diversity of animal species, it is only inferior to Altai.

On the territory of national park there are monuments of history and culture, which are objects of ecological tourism. This is the Uygentas burial ground, located in the Uygentas tract on the right bank of Tentek river, burial mounds in the Sarymsakty river valley, on north-western slopes of the Karaagash ridge, on the Koktobe pass (1814 m), in the Runjurek mountains, and also the Dinar burial ground of Khorezm shah Mohammed ben Tekesha, dating from the thirteenth century.

The *Buiratau state national park* created in 2011 is particularly interesting from the ecotourism point of view. On the whole park territory, quite unique steppe ecosystems are presented that belong to the dry steppes type. At the same time, its relic black-alder forests in birch groves are considered to be the property of Buiratau park. Entire territory of the Buiratau Park lies in zone of the Central Kazakhstan hillocky area. The name of the park in many ways characterizes its nature: the word "Buiratau" translates from Kazakh as "curly mountains". Local natural ecosystems do abound with birch and alder forests (Fig. 2.45).

Total number of vegetation, about 30 species are classified as very rare and even endangered, since their habitats are gradually decreasing. It is a sticky alder, spring Adonis, open chamber, Siberian skerda, barn fern, fragile bladder, and plants such as steppe peony, raven eye, Elbe woods and Gesner tulip.

Fig. 2.45 The Buiratau state national park (*Resource* Available at https://Kazakhstan.travel/photobank)

Avifauna of the national park deserves special attention, in which 227 bird species are registered, and from them 127 are nesting species (Tourism of Kazakhstan 2015).

The natural reserves have great interest for ecological tourism. Today, there are five reserves in Kazakhstan.

The Irgiz-Turgai State The natural Reserve was created in order to preserve the natural complexes and representatives of the Betpakdala population such as saigas. The reserve is a place of concentration of migratory waterbirds. The Irgiz-Turgai reserve occupies the flat areas of Kazakhstan, located in the semi-desert and desert landscape areas of the temperate zone of the Caspian-Turanian country. Its main part belongs to the Turgai-Central-Kazakhstan region.

The saiga lives in the southern part of the reserve, it is a representative of the Betpak-dala population.

Saiga (Saiga tatarica) is a migratory artiodactyl that forms five main populations, three of which are in Kazakhstan, one in Russia (Kalmykia) and one in Mongolia. The Betpak-dala saiga population was and remains the largest among all populations. Territory of protected areas is a place of historical settlement, habitat, wintering, summering of the saiga.

The state nature reserve Akzhayik is located in Atyrau city and Makhambet district of the Atyrau region. Relief of the Caspian lowland within reserve is formed mainly by the sea activities. Main features of the relief are associated with the general geological structure of territory. Reserve area is characterized by flat terrain, weak surface dissection and low relative elevations.

The following types of relief are distinguished: marine accumulative plain, alluvial-delta plain, and fragmentary plain of denudation origin. The delta of the Ural river with adjacent water area of the Caspian Sea is a land of international importance. The rich feeds of extensive reservoirs attract a lot of waterfowl and wetland birds. Avifauna of the nature reserve is very rich and interesting. In delta of the Ural river and on adjacent coast of the sea 292 bird species are recorded. Total number of birds in the migration period, according to expert estimates, reaches 3 million individuals.

Delta of the Ural river and adjacent water area of the Caspian Sea ensures the existence of rare species large number concerning water and near-water birds. Some of them breed in this area—the curly pelican, small cormorant, the Egyptian heron, little white heron, whooper swan, yellow heron, sultan, black-headed laughter.

The reserve contains 48 species of mammals belonging to 7 orders, which makes up more than 25% of the mammalian species list in Kazakhstan. The Caspian Sea is also a unique inhabitant of territory.

In delta of the Ural river with adjacent coast of the Caspian Sea 229 species of higher plants are registered, belonging to 141 genus, 56 families (Biosphere Reserve Ak-Zhaiyk https://www.kazmab.kz/index.php/biosfernye-rezervaty1/natsionalnaya-set/ak-zhajyk/opisanie).

Schrenk's tulip is among the "red book" plants. Obviously, the tulip species has spread from more northern steppe territories, where it is common.

Nut of Kazakhstan. It is an endem of Kazakhstan, and can be found in delta of the Ural river with adjacent coast of the Caspian Sea in low-flowing water bodies, sometimes covering the entire water surface.

Salvinia floating is water fern and macrophyte. It floats on the surface of standing and slowly flowing waters, forming dense thickets, blocking the access of light and thereby changing environmental conditions of water bodies. Salvinia is present along with Kazakhstan nuts in the reserve ecosystems. Salvinia thickets are a good haven for young fish.

The state institution called Semey ormany state forest the natural reserve is located in the East Kazakhstan.

Total area of the Semey ormany reserve is 665,502 ha, including forest land, which is 392,802 ha.

Reserve has protected the magnificent ribbon bores, formed on ancient sand deposits, formed in the past between the Ob and Irtysh rivers.

The ribbon burs of Kazakhstan belong to the Semipalatinsk-Srostinsk borovoe district.

The state institution called *Altyn-Dala state nature reserve* was established in November 2012 on territory of the Amangeldy and Zhangeldy districts in Kostanay region with a total area of 489,766 ha, including the territory of Sarykopinsk state nature reserve of republican importance in order to preserve saigas and steppe ecosystems.

The reserve territory is located in the South-Turgai physiographic province of the West-Siberian Kazakhstan region in Eurasian steppe zone between the Mugalzhars in west and the Ulutau mountains in east. From south it is bounded by the vast hollow Chelkar-Teniz.

Its territory includes the main habitats, migrations, and lambing of the world's largest Betpak-dala saiga population. Functioning of the Altyn Dala reserve provides for year-round stationary protection of the saigas population and contributes to the increase in their numbers (Fig. 2.46).

Reserve flora has more than 370 species, among which 23 species are assigned to a different category of rarity. Here you can also find rare plant communities for Kazakhstan–tyrsik steppes and turang groves.

It also presents the largest closed river basins of Kazakhstan (the Turgai and Uly-Zhylanshik rivers) and unique lake systems, including the region's largest freshwater system of lakes Sarykopa—a key ornithological area. The reserve's territory is also a pilot site for implementation of the Przhevalsky horse and cooler reintroduction program, which are listed in the International Red Book.

The republican state forest *the natural reserve Ertis ormany* is located in the ribbon forests of Pavlodar region on the Irtysh right bank. It is on territory of the Shcherbakty area and Lebyazhinsk area of the Pavlodar region.

The object of eco-tourism is pine "Orman batyry", growing in the protected core centre. Butt and lower part of the boot on one side are damaged by fire and have signs of tapping. This pine is the oldest and largest tree in reserve and still vegetates satisfactorily and forms annual shoots.

Fig. 2.46 Saigas (*Resource* Available at https://Kazakhstan.travel/photobank)

Following the reserve creation, staff began to study the species diversity of animal world.

The Republic of Kazakhstan, possessing significant tourism and resource potential, has great opportunities for development of tourism at the international level. Moreover, in conditions of the continuing deterioration on ecological situation, the development of ecological tourism is simply necessary.

Ecotourism pursues the goals of preserving the natural environment in it's the natural state and at the same time brings a certain profit, meeting the interests of residents.

References

Aliyeva Zh N (2001) Rol ekologicheskogo turizma v ustoychivom razvitii Respubliki Kazakhstan (na primere Almatinskoy oblasti): Dis. kand. geograf. nauk. (Role of ecological tourism in the sustainable development of the Republic of Kazakhstan (on example of the Almaty region): Diss. Cand. of Geography Sc.). Almaty

Aliyeva ZhN (2002) Ekologicheskiy turizm (Ecological tourism). Kazakh University, Almaty, p 101

Aliyeva ZhN (2019) Ekologiyalyk turizm (Ecological tourism). Kazakh University, Almaty, p 115

Artemyev A. The Ertis River, image

Artemyev A. Birch scraps of Ulytau, image

Artemyev A. Kazakh hillocky area, image

Artemyev A. Lake Issyk, image

Artemyev A. Glacier Tuyuksu, image
Artemyev A. The Almaty reserve, image
Artemyev A. "Ile-Alatau" national park, image
Artemyev A. The Zhoshy Khan mausoleum, image
Biosfernyy rezervat Ak-Zhayyk (Biosphere Reserve Ak-Zhaiyk). https://www.kazmab.kz/index.php/biosfernye-rezervaty1/natsionalnaya-set/ak-zhajyk/opisanie
Kazakhstan.travel https://Kazakhstan.travel/photobank
Severnyy Kazakhstan. O regionakh Kazakhstana. Turizm regionov Kazakhstana (North Kazakhstan. About the regions of Kazakhstan. Tourism of the regions of Kazakhstan). https://www.orexca.com/rus/Kazakhstan/regions/northern_Kazakhstan.htm
The National Atlas of the Republic of Kazakhstan (2010) The natural conditions and resources. vol. 1, Almaty, 150p
Tsentralnyy Kazakhstan. O regionakh Kazakhstana. Turizm regionov Kazakhstana (Central Kazakhstan. About the regions of Kazakhstan. Tourism of the regions of Kazakhstan). https://www.orexca.com/rus/Kazakhstan/regions/central_Kazakhstan.htm
Turizm Kazakhstana (Tourism of Kazakhstan) (2015). Bastau, Almaty, p 520
Vostochnyy Kazakhstan. O regionakh Kazakhstana. Turizm regionov Kazakhstana (Eastern Kazakhstan. About the regions of Kazakhstan. Tourism of the regions of Kazakhstan). https://www.orexca.com/rus/Kazakhstan/regions/eastern_Kazakhstan.htm
Yerdavletov SR (2010) Geografiya turizma (Tourism geography). Kazakh University, Almaty, p 412
Yerdavletov SR (2015) Geografiya turizma Kazakhstana (Geography of tourism in Kazakhstan). Kazakh University, Almaty, p 180
Yuzhnyy Kazakhstan. O regionakh Kazakhstana. Turizm regionov Kazakhstana (Southern Kazakhstan. About the regions of Kazakhstan. Tourism of the regions of Kazakhstan). https://www.orexca.com/rus/Kazakhstan/regions/southern_Kazakhstan.htm

3 Social and Economic Tourism Potential of Kazakhstan

3.1 Socio-economic Recreational Resources

Socio-economic recreational resources, or otherwise they are called cultural and historical resources, are one of the most important factors determining the degree of tourist attractiveness of particular territory. The areas long inhabited by human being keep history, traditions, and culture. Cultural and historical recreational resources are prerequisites for the organization of cultural and educational types on recreational activities, on this basis, they optimize recreational activities in general.

Among cultural and historical monuments, the leading role is played by historical and cultural ones, which are the most attractive. The typology is based on informational essence of cultural-historical object: uniqueness, typicalness amid objects of this kind, cognitive and educational value, and visual appeal.

On the territory of Kazakhstan there are a significant number of monuments reflecting the history of country from ancient times to present day. Most of them are archaeological monuments, which include settlements, burial mounds, burial grounds, petroglyphs, ancient irrigation structures, ancient mines, and stone sculptures.

West Kazakhstan. West Kazakhstan can be considered as very interesting in terms of ecotourism. Being at the junction of two world parts—Europe and Asia—territory of Western Kazakhstan was constantly influenced by various ethnocultural entities. Territory was crossed by numerous trade routes and hordes of Mongol conquerors. Mineral riches of the Mugodzhar mountains, unique the natural complexes of the Ural (Zhayik), Emba (Zhem), Irgiz and their tributaries, and oasis of sandy massifs on the region was attracted by ancient people. West Kazakhstan, making up a vast territory, occupies one of the important places in history on formation of the Kazakh statehood and the Kazakh ethnogenesis. Big advantage played its diverse climatic conditions that ensured the development of nomadic and, later, agricultural crops. These factors contributed to the fact that numerous historical and cultural monuments of interest to the ecological tourism development have been preserved in this area.

K. Iskakova et al., *Ecological Tourism in the Republic of Kazakhstan*,
https://doi.org/10.1007/978-3-030-77462-2_3

Ancient sites of the Mugodzhar mountains, the Emba river basin, and the Ustyurt plateau may be of great interest for tourists. It was established that the main occupation of Ustyurt plateau population in the Neolithic age was hunting with frequent migrations from one camp to another. Apparently, this is due to the fact that this region was characterized by favourable climatic conditions, an abundance of feed for livestock, and presence of hunting objects. Stone Age sites were also found on the right bank of river Ilek (near the regional centre), and group of sites—on the river Irgiz.

It is established that most of the archaeological objects in north region belong to Sarmatian culture of the early Iron age. For the most part, these are barrows left by Iranian-speaking nomads who lived in the steppes of Southern Urals in period of the VIII–IV centuries BC. This object reflects the interaction of Pechenegs, Oguz and other tribe cultures of that time. There are various historical and cultural monuments in this territory, the region also has significant recreational resources (Urkachevsk relic forest, springs, etc.), which make it possible to organize ecotours.

Koskuduk site in the Mangistau region, dating from the Eneolithic (late Stone age), is located 7 km north of Aktau. Here were found the remains of stone walls in residential buildings, homeowners, venues of ancient people religious rituals who inhabited Mangystau 5 thousand years ago BC.

Necropolis Kyrykoba. From VII century BC steppe regions of West Kazakhstan were inhabited by tribes known in written sources under the general collective name Savromats, and from the end of fourth century BC—Sarmatians. Among the excavated Savromat monuments of West Kazakhstan, the most striking materials were obtained from the Kyrykoba kurgan necropolis, where graves of tribal leaders, military aristocracy and priests of the early nomads in West Kazakhstan and the Southern Ural region were buried.

Monument is located on the territory of Burlinsk district in West Kazakhstan region. The monument complex consists of two parts: the southern part consists of two burial complexes consisting of mounds, two parallel shaft-like structures adjacent to them.

The settlement Saraichik is located 50 km from the city of Atyrau on the right bank of the Zhaiyk River. It is the capital of the first Kazakh state formation, created by Khan Kasym. According to archaeological findings, in particular, vessels with inscriptions, Saraichik in the XII century, was a city with a developed culture, trade and craft. Arab merchant-traveler Ibn Battuta, who visited these places, described his stay in 1333 in Saraichik, as a city which connected the East with the West (Tourism of Kazakhstan 2015) (Fig. 3.1).

To the south of Uralsk, there is the medieval settlement of Zhaiyk, which includes residential buildings, as well as industrial and public buildings, such as an oriental bath, brick kilns and the remains of ceramic kilns. City on the Zhaiyk settlement site appears in XIV century during the reign of the Uzbek khan, and Zhanibek, characterized in the Golden Horde history as years of the greatest stability and flourishing of urban culture. Settlement of Zhaiyk as a monument to urban culture of the Zhoshy Khan period conveys perfectly the medieval appearance of the city, its life, culture and household economy.

Fig. 3.1 Saraishyk settlement (*Resource* Available at https://Kazakhstan.travel/photobank)

Thus, the territory of West Kazakhstan is truly an open-air museum, the brightest exhibits of which are stored in Mugalzhar, Ustyurt, Priaralye, Ural, etc. Organization of complex expeditions to study historical and cultural monuments would contribute to the development of educational value.

For the development of ecological tourism, underground cult complexes of West Kazakhstan, the so-called underground mosque complexes, are noteworthy. The ancient necropolis, according to folk lore, arose and expanded around the tombs or underground mosques of the first Islam preachers. There are plenty underground mosques cut down in the massif of pre-rock cliffs and on the mountain slopes, on Ustyurt and Mangystau. They are Shopan-ata, Beket-ata, Shakpak-ata, etc. (Fig. 3.2).

Shakpak-ata mosque is carved in the rocky cape thickness formed by the two ravine beds. Beket-ata is legendary leader of the Aday family in Kazakh, he built several underground structures, one of which is the Beket-ata complex in Oglandy. Underground mosque Beket-ata—the centre of pilgrimage in West Kazakhstan.

Shopan-Ata mosque is the largest, and possibly the oldest, in Mangistau. It is located on both sides of a shallow and narrow ravine, on the slope of Shopan-ata mountain.

Fig. 3.2 Underground mosque Beket-ata (*Resource* Available at https://Kazakhstan.travel/photobank)

All monuments of the region are concentrated in ancient necropolises and cemeteries dating from the XV–XX centuries. The most well-known monument of Aktobe region is the mausoleum of Abat-Baitak, which is known as an outstanding monument of religious architecture in Kazakhstan (Fig. 3.3).

There is a great interest for ecotourism in the “Khansk burial ground” (Khan molasy), located near the confluence on Ulkayak and Kabyrga rivers. The object is remarkable by the fact that here, according to sources and stories of old-timers, Khan of Junior zhuz—Abulkhair is buried.

Among the historical monuments of Mangystau a special place is occupied by the city—museum Fort–Shevchenko. In the Middle Ages, a commercial quay was located here—the fortified settlement of Ketikkala, which existed in the X–XV centuries. The mound had economic ties with shopping centres in Europe and Central Asia, as evidenced by archaeological finds on its territory. During Russian King Peter I, there was the first Russian fortress of St. Peter. According to the staff project—captain Ivanin, the impregnable Novopetrov fortress was erected here. The settlement near fortress was renamed Fort–Aleksandrovsk, in the first years of Soviet power it was called Fort–Uritsk. Since 1939, the city was named Fort–Shevchenko. On the city outskirts of mountains Korgantas the ruined remains of fortress are preserved.

Fig. 3.3 The Abat-Baitak mausoleum (authors)

Central Kazakhstan. Immense expanses of Central Kazakhstan from ancient times were the scene of many historical events that left an indelible mark on cultural and economic development of the tribes and people living here, which to some extent was reflected in the monuments preserved in region.

Central Kazakhstan was one of the main areas on establishment and development of the Andronov tribes. In the Karaganda region, settlements of Andronov and numerous kurgans dating from the Bronze Age and Early nomads were discovered.

The Begazy-dandybay archeological culture of Bronze Age is named after the burial grounds of Begazy and Dandybai in Karaganda region (Fig. 3.4).

The Begazy burial ground is considered as one of the largest memorial structures, belonging to the Bronze Age. The complex includes a large number of burial grounds on different time and character. For the construction were used granite slabs of granite weighing up to three tons. There is a great interest on supra-gravestical structures made of stone. The Begazy monuments represent one or two-chamber buildings of square shape, the sides of which are strictly oriented to cardinal points. This similarity is expressed in coincidence of the main structure plans and partly in wall design, made of horizontal Stone Age (Tourism of Kazakhstan 2015). According to scientists, this period is defined as the transition from the Bronze Age to Iron Age.

Surviving remnants of sedentary culture, the most significant is site of ancient settlement Dombauyl in mountains of Arganaty, Ulytau district. There are remains of garden, arable land, melon fields, as well as an irrigation network. In the Ulytau steppe, still there are some examples of irrigation system, hydrotechnical and

Fig. 3.4 The Begazy burial ground (*Resource* Available at https://Kazakhstan.travel/photobank)

irrigation structures in the form of numerous dams (togan), and wells. The Kipchaks irrigated fields not only with water brought out of the rivers, but also from the reservoirs and lakes by means of the Chigyrly construction and water-lifting machines. In steppe, one of the rivers flowing from the Ulytau mountains got its name Chigyrly from the word Chigyr, thanks to the fact that it irrigated a large plain and there were many water-lifting devices on its coast.

Initial localization of the Central Kazakhstan sites in VIII–XI centuries linked to history of the Oguz tribes. This is confirmed by the archaeological complex of sites belonging to the Oguz and Pechenegs. Baskamyr settlement in the natural boundary Taldysai of Ulytau district is the most significant among the other citadel settlements dated by Oguz-Pecheneg period (Fig. 3.5).

The Baskamyr settlement is located 83 km northwest of the city of Zhezkazgan in the Karaganda region. These are the remains of a medieval castle, enclosed by three powerful walls with citadel and flanking towers. The length of shaft is 108–110 m, and width is 14 m. The remains of defensive tower in a powerful hill formed in places up to 5 m (Tourism of Kazakhstan 2015). This vast territory was divided by Chinghis Khan between sons. The eldest son Zhoshy was assigned to the steppes of modern Kazakhstan. In connection with formation of the Zhoshy ulus and the location of its headquarters in foothills of the Ulytau, these mountains were called the "great mountains" (Ulytau). It is located on the Karakengir river left bank (Fig. 3.6).

Fig. 3.5 The settlement Baskamyr (*Resource* Available at https://Kazakhstan.travel/photobank)

Fig. 3.6 The Zhoshy Khan mausoleum (photo taken by Artemyev A.)

Fig. 3.7 The Alashakhan mausoleum (*Resource* Available at https://Kazakhstan.travel/photobank)

Zhoshy Khan mausoleum is a monument of architecture in the XII–XIII centuries, and also located on the Karakengir river left bank. The mausoleum was erected from a good burnt brick, covered with blue azure. The sculptural top, which resembles a head with horns, attracts attention in the mausoleum. It is made from pottery clay by burning. General form of the mausoleum: a cube with raised front wall—peshtak, architecturally decorated entrance. The height of the peshtak reaches eight meters. The mausoleum makes a very grand impression, especially after resurrection. Among the Ulytau settlements, the most ancient is the town of Alashakhan. These are the remains of huge ramparts and defensive walls, built of large raw brick and clay, mixed with gravel and river pebbles (Fig. 3.7).

Smooth outer surfaces of the side and rear walls of bases have ornamental brickwork laid in the form of triangles, rhombuses, Christmas trees, etc. The combinations of such pattern motifs give the impression of carpet. The entrances to mausoleum are decorated in form of niches, which is a typical feature of cult architecture among the Kazakh in tenth–fifteenth centuries. By its construction methods and stylistic peculiarities, the Alashakhan mausoleum is the only structure of this type for both territory of Kazakhstan and whole Central Asia. In Central Kazakhstan, the remains of multiple settlements, castles, fortifications were found: in the Nury basin, Auliebulak near the sanatorium Zhosaly, remnants of settlements along the Sherubay-Nura river, on Sarysu, in the Kengir valley, Zhezdy, and in Karkaraly. The most interesting remnants of ancient settlements are located in the gorges and foothills of the Ulytau and adjacent to the upper reaches of rivers Turgai, Kengir, Zhilanshyk, Zhezdy. Judging by them, in the early Middle Ages and later this area was a significant centre of sedentary culture and trade, through which the caravan routes passed, connecting with sedentary agricultural oasis in the river Syrdarya.

North Kazakhstan. North Kazakhstan is the main granary of country and one of the important areas of grain farming, meat and dairy farming. Thanks to millions of hectares raised virgin soil, Kazakhstan has now become one of the main suppliers of "hard" grain to the world market. And thanks to the development of Kostanay iron ore basin with high-quality magnetite ores, it became the largest supplier of iron ore to the plants of Ural region in Russia.

Not only Kazakhstan people raised the virgin lands in the republic, representatives from all republics and regions of the former Soviet Union visited here. Currently, there is a kind of pilgrimage. Former virgin lands with families are travelling to Kazakhstan to show their relatives where the first tent was, where the first furrow was laid, and then they gathered a rich harvest.

There is a particular interest from ecotourists to North Kazakhstan. The culture of North Kazakhstan tribes has much in common with the Sak tribes culture of southern Kazakhstan. Some scientists believe that mounds in the Ulybay area in North Kazakhstan are similar to the monuments of Tasmola culture. Monuments of early iron in North Kazakhstan make up the Ulybay-Tasmola culture.

Monuments of the early Iron Age in Central Kazakhstan are represented by burial mounds. This region is characterized by so-called barrows with a "moustache". According to some scientists, they played the role of lighthouse in steppe, as they were built at a certain angle. Monuments of the early Iron Age of Saryarka belong to Tasmola culture. Name is given by the Tasmola locality in Ekibastuz district of Pavlodar region, where the mound of this culture was excavated for the first time. "Moustache"—stone chains, always moved away from the east side of mound. Apparently, this testifies to the Tasmola worship of the Sun. Moustache mounds are the most frequently encountered monuments of early Iron Age in Central Kazakhstan.

The Botay settlement. One of the most interesting objects of ecological tourism is the oldest settlement of Botai. Scientists managed to reconstruct the entire process of housing construction by ancient builders. Initially, they dug out a polygonal or round pit. The depth of pits 60–80 cm, the area from 30 to 70 m^2. Then they drenched the clay and put them in lumps into walls, which from the inside and outside were reinforced with animal bones. The width of walls on ancient dwellings was 80–120 cm, height—from 60 to 100 cm. Special pits were dug near the dwellings, from which they used clay for coating walls and floors. Tent ceiling with a hole in the centre for smoke was erected around the walls from its perimeter. The gaps between logs were plastered with clay, and on top they were covered with turf and animal skins. Thus, the height inside room was 250–320 cm. As it turned out during the excavations, in ancient times the living quarters of Botay often closely adjoined each other. The following can be said about internal planning of dwellings. In the centre on floor was the hearth. As a rule, there were small household pits around the walls. Within the walls of pits, numerous niches were made for household and religious purposes. The booths slept on mats, hides, or made low planks along wall opposite to the exit. There were interesting "pits-canned", which kept the meat for a long time. This was done as follows: they dug a hole to depth of 1 m into which parts of the horse's carcass were put, then covered with hides and clay. Fire was made from above, which burned until all the oxygen was burned out. After cooking, the meat could be there for a long time without access to air. The earliest of all known evidence on horse domestication has been found on the Botay (Tourism of Kazakhstan 2015). There are numerous Andronov monuments of interest to ecotourists in North Kazakhstan. Many archaeological sites represent all periods of human history.

Settlement and burial ground Bestamak. The necropolis is located in north part of the Turgai trough at the very source of the Ubagan river, the right tributary of the Tobol in Kostanay region. Total area of the excavation is currently 9092 m^2. During the period of research, more than 170 burial pits and ritual structures were opened and studied at the burial ground. The Bestamak is unique place, the materials of which are represented by wide archaeological framework: from the Eneolithic to the Middle Ages.

Interesting settlements and burial grounds of the Bronze Age, mounds of early and medieval nomads, located along banks of the Irtysh river. Archaeological monument "Bekteniz" includes a lot of mounds. Monument name was the ruins of a small village. Most of the excavated mounds belong to the Bronze Age. For the first time in archeology of North Kazakhstan, an original complex was obtained, containing bronze mirrors, altars, and women's jewellery. On the territory, archaeologists have found a burial field, where the burials of people were made belonging to a special social group—the priests.

Mirror in ancient times was an object with magical fairy powers. It served as an obligatory attribute of Sarmatian priestesses. The mirrors were placed in grave with stone altars and ornaments. Sarmatians lowered them into the grave deliberately broken. Some of them were exposed to fire.

In Arshalynsk district of Akmola region there were found a complex of mounds "with a moustache" of early nomads in the Middle Ages. In village of the Korgalzhyn is the Bytogay mausoleum, in the Zharkainsk district—the Kyzyltam mausoleum, in Zhaksynsk district—the Baubek batyr mausoleum. The kurgan groups "Kyzyl-su" are located in the Esil district.

Extremely dense saturation of archaeological monuments in the Akmola region is a remarkable phenomenon, creating prerequisites for ecotourism development. Almost throughout the territory of Akmola region there are amazing monuments of religious architecture—mazars, revered by Muslims as holy places and representing buildings over the graves. They were built from raw bricks. Despite the rather primitive methods of construction and materials, some monuments have stood the test of time.

Currently, there are plenty of architectural monuments of the XII–XIX centuries only in territory of Akmola region: Botagay mausoleum (XI–XII centuries), Bespakyr mausoleum (XIX century), mazar by Aliptomar master (XVIII–XIX

Fig. 3.8 Karasai and Agyntai heroes memorial complex (authors)

centuries), mazar Kanykei (XVIII century), mazar Barshingul (XIII century), mazar Ali Tusup (XVII centuries) rise above the expanses of Ishim, surprising by the forms originality. Hundreds of pilgrims, despite their great distance, come here to heal.

Karasai and Agyntai heroes' memorial complex is located near the village of Madeniet in the North Kazakhstan region. Monument in honour of the well-known heroes and fighters for independence of the Kazakh Khanate as Karasai and Agyntai was opened on October 15, 1999. The memorial complex of Karasai and Agyntai heroes is an architectural monument. Compositional complex figuratively develops the traditions of ancient Turkic kagans' stakes (Fig. 3.8).

The whole complex is focused on its main axis to Mecca, where the ledge of the mosque mihrab is directed. Entrance to the circle enclosed by a brick fence is highlighted by two towers and leads to platform in front of the mausoleums intended for prayer and sacrifice. At the site corners, on granite bases, as symbols of military valour, crossed spears are set from a metal profile with bronze overlays. In the centre of circle there are two mausoleums of heroes, which have ellipsoidal outlines as characteristic of North Kazakhstan. In front of them is a small cone-shaped memorial mosque, symbolically and architecturally uniting mausoleums. In the centre, between mausoleums, there is a small pinakothek-hall, where texts in three languages engraved on walls, telling about the heroic deeds of the batyrs (Tourism of Kazakhstan 2015).

Complex residence of Abylay Khan. Abylai Khan's residence is located in city of Petropavlovsk, North Kazakhstan region. Archival documents confirm the construction of Abylay residence in autumn of 1765 year. This is a historic complex, which includes office, the Khan's chambers, and guest house.

Confirmation by archival documents of the Abylay residence construction in autumn of 1765 is an example of the distant events' reconstruction which are in particular importance for the objective history of Kazakhstan. These works were carried out within the framework of the first State program stage called "Cultural heritage".

Construction work lasted two years, and in August 2007, grand opening of the historical-memorial complex "Abylay Khan residence" took place in Petropavlovsk (Tourism of Kazakhstan 2015) (Fig. 3.9).

East Kazakhstan. East Kazakhstan occupies a tenth place in Kazakhstan area. Its territory is the smallest among economic regions of Kazakhstan. It is located in the upper reaches of the Yertis river—the largest waterway of the republic. Yertis cuts through the territory of the region from the south-east to north-west and connects it with West Siberia. Railway Lokot–Semey—Almaty gives East Kazakhstan an exit to West Siberia, and through South Kazakhstan region to Central Asia.

Fig. 3.9 Complex residence of Abylay Khan (authors)

East Kazakhstan is the most important in the republic and one of the largest in CIS regions on non-ferrous metallurgy and the hydropower production. Prominent place in the economy of district is also occupied by agriculture, light industry and food industry developed on its basis.

Burial ground Shilikty. Shilikty barrows are located in the valley of the same name in the Zaisan district of East Kazakhstan region. On three sides, the valley stretching 80 times in length and 30 km wide is surrounded by mountain ranges: in the south and west—Tarbagatai, in the east—Sauyr, in the north—Manyrak (Bayandinova et al. 2018). In the Shilikty valley there are more than two hundred monuments of early Iron age, which is about 130 mounds of Sak-Scythian time are located in its central part, on area covering 1.5 km wide and 6 km long. Such a dense concentration of elite monuments on Sak culture in small area of Kazakhstan is very rare. In addition, in the valley and foothills there are settlements and burial grounds of the Bronze Age, fences, mounds and stone sculptures of Turkic time. Mountains surrounding the valley are rich in rock paintings dating back to period from the Bronze Age to late Middle Ages. Sensational material was obtained from the Baighetobe mound, where the third golden man in Sak was discovered on the territory of Kazakhstan. The total of 4303 gold items were found in the Baytetobe burial mound of Shilikty – 3, among which: 153 badges in the form of leopard mask, 36 badges in the form of golden eagle, 20 badges in the form of deer, 39 badges in the form of wolf cub (bear), 17 gold leaf strips, 7 pieces of gold leaf wires, 141 miniature pendants of hemispherical shape, 2835 miniature cup-shaped pendants with soldered ear, 223 tiny tubular beads, 743 pieces of ring-shaped beads and tiny hook. Mound Baytetobe from the group Shilikty-3 was grandiose architectural structure. The mound size as a social sign does not cause any doubt that one of the Sak kings was buried here (Tourism of Kazakhstan 2015).

Ruins of the Ablayket temple. Ablayket (Ablainkit)—Dzhungarian fortified Buddhist monastery of the XVII century. It was founded in 1654 by taishi Ablai. In 1671, during the internecine struggle, Galdan was taken and doomed to desolation. The monastery ruins are located on territory of the Ulan district in East Kazakhstan region. Complex was located in mountains and in plan had the shape of pentagon. Along the perimeter it was surrounded by wall up to 2 m high. The appearance of such temples is explained by policy of mastering in the seventeenth century, which were captured land by the Dzungars and establishing their control over the emerging trade routes from Russia to China. The walls were protected by two religious buildings, which in the XVIII century manuscripts in Mongolian language, Buddha statues were discovered (Tourism of Kazakhstan 2015).

From the middle of XV century, the Irtysh area began to be settled by the Kazakh, who created their own khanate here. In the XVI century, various conquerors invade here, in 1643 the Dzungarian Khan Batur conquered almost the entire valley of the upper and middle Yertis. To strengthen his power, he built monasteries on the Yertis shores, including the largest—Zardzhinkit (1648), which consisted of seven huge clay buildings (Seven chambers). In 1670 the monastery was destroyed during feudal wars of the Dzungars, but its ruins remained for a long time. According to them, the fortress was later named, and then the city also (Yerdavletov 2015).

Semipalatinsk fortress was founded in 1718, under Peter I, 18 km from the present city. In its place is now the village of old Semipalatinsk. In 1778, near the ruins of Seven chambers, Russian soldiers built a new fortress, which quickly began to turn into a city (Bayandinova et al. 2018a, b).

Before the 1917 revolution, Semipalatinsk was the administrative and commercial centre of vast region. Large merchants were the main owners of it, and they profited from the deception and exploitation of the Kazakh population, economically to Semipalatinsk (Yerdavletov 2015).

For a long time the emblem of old Semipalatinsk was decorated with a golden camel, emphasizing the city importance as a major trading centre on Yertis (Bayandinova et al. 2018a, b). But it can be said that at the same time Semipalatinsk was the gate that opened way to the heart of little-studied Asia. That is why it attracted the attention of travelers and researchers such as P.P. Semenov-Tien-Shan, G.N. Potanin, Ch. Valikhanov and others.

F.M. Dostoevsky in middle of the last century served a term in this city. After a convict prison in Omsk, a soldier's barracks in the Semipalatinsk seventh line battalion, he can live in a private flat. Dostoevsky settled in the poor hut of widow-soldier in the Russian part of city, here for the first time in recent years he taken a pen. The first etudes and sketches of future "Notes" are born, which soon brought him worldwide fame. Here he created his novels "Uncle's Dream" and "The village of Stepanchikovo and its inhabitants".

Always crowded in rooms of the literary-memorial museum of F.M. Dostoevsky, there is Dostoevsky street and a library named after him in Semey.

Abai Kunanbayev studied and periodically lived in Semipalatinsk. The best years of his life and work are connected with this city. On one of the streets is a small wooden house—the memorial museum of great poet. There is a monument to Abai in the street named after him. Mukhtar Auezov learned to write in this city (Yerdavletov 2015).

Before the revolution, city was often called the "devil's sandbox", because sands were attacking the city from three sides. Now Semey is a very green city and here you can find a saving shadow everywhere on hot summer days. Townspeople love to relax on the former Colonel island, where a beautiful park is now laid out, or in the ribbon forests of the Yertis right bank. Only half an hour away by car separated the centre of city from the edge of red-stemmed pines, with peace and quiet, broken only by the rustling on dry pine needles and the sound of wind in tops of trees.

Architectural monument of the city in 1776 is the fortress gate. Not far from the city are the ruins of Semipalatinsk fortress, which gave the city name.

There are other monuments in the area. This is a monument to Enlik and Kebek (Kazakh Juliet and Romeo). From distance, a tall pyramidal obelisk is visible, mounted on massive foundation of hip form. Inside are two graves, on them is a plate with the inscription: "Enlik-Kebek. The middle of XVIII century. Tribal feudal lords were executed for protesting against feudalism and powerlessness of

Kazakh women." Behind the dry lines of museum text there is a story about the great tragic love of Kazakh girl Enlik and horseman Kebek, described in his time by Abai (Tourism of Kazakhstan 2015) (Fig. 3.10).

In the tract Zhidebay, where the poet lived for many years, Abai's wintering was turned into a branch of his Semipalatinsk memorial museum. Not far from the museum is the Abay tomb.

Mazar of Kozy Korpesh and Bayan Sulu. It was erected near the village Tansyk in East Kazakhstan region. It is one of the oldest monuments in Kazakhstan, which have reached our time. Total height of the mazar is 11.65 m. Chokan Valikhanov, who visited the monument in 1856, was struck by its height. The monument is associated with the ancient legend of Kozy Korpesh and Bayan Sulu tragic love. The monument serves as a symbol of this immortal love (Fig. 3.11).

In front of the entrance to the mausoleum, four sculptures are installed in pairs, which, according to popular legend, are depicted by Kozy Korpesh, Bayan Sulu, her younger sister and daughter-in-law.

In 1856, the mausoleum and sculptures next to it were painted by Chokan Valikhanov. In 1858, mausoleum was surveyed by Nikolai Abramov, and in 1898 —by Nikolai Pantusov.

In 1952, the monument was investigated by an archaeological expedition of the Kazakh SSR Academy of Sciences under Alkei Margulan direction.

In 1982, the mazar of Kozy Korpesh and Bayan Sulu included a list of historical and cultural monuments of the Kazakh SSR on national importance and was taken under state protection (Tourism of Kazakhstan 2015).

South Kazakhstan region (Turkestan region). South Kazakhstan region includes a vast territory from the Aral Sea in west to the Zhongar gate in the east and from lake Balkash and the Betpak-dala desert in the north to the border of republic in the

Fig. 3.10 Monument of Enlik and Kebek (own authors)

Fig. 3.11 Mazar of Kozy Korpesh and Bayan Sulu (*Resource* Available at https://Kazakhstan.travel/photobank)

south. Diversity of socio-economic recreational resources creates favourable conditions for the ecotourism development.

Ancient Talgar city. The Persian geographer of late tenth century composed the geographical treatise “Hudud al-Alem” (“Borders of the world”) in 982. On its pages, among the listed settlements of Semirechie, village of Talkhiz was named. It was located among the mountains on border between the Turkic tribes of Chigily and Karluk. “Its inhabitants, according to the characteristics of medieval geographer, are belligerent, bold, and valiant”.

Analyzing the ancient name, scientists came to conclusion that it was distorted by the copyist of original work and sounded differently—Talhir. Talhir (Talgar) is a well-known name: this is name of the highest peak in Zailiyskiy Alatau (now Talgar peak), a swift mountain river and modern city. After such an assumption it remained to find the remains of medieval city, and it was found. The predecessor of modern Talgar was located on the Talgar river right bank at its exit from mountain gorge into valley. Now, no one from scientists has any doubt that this particular settlement corresponds to settlement of the tenth century Talhiz–Talhir. The ruins of Talgar are now a sublime quadrangular area, surrounded on all sides by swollen shaft. Area occupied by the fortified part is about 9 ha. Towers were located at the corners of shaft and around perimeter. Behind the shaft even now you can see the remains of moat. Two entrances were opposite each other in the middle of the south-west and north-east walls. They were connected by road that divided the entire territory of city into two almost equal parts. Currently, archaeologists opened cobblestone pavement, which was the main street of city. The fortified central part of city was adjoined from all sides by buildings that constituted a kind of trade and craft suburb (rabad). Together with them, the city occupied a territory of nearly 30 ha. In order to determine the city time, periods of ups and downs in its life, the cause of death, find out who inhabited the city, what citizens did, how they lived, archaeological excavations were carried out on the settlement, etc. In the second half of XIX century, the city of Talgar was visited and prominent travelers and scholars of that time visited and were keenly interested in its history: Semenov-Tien-Shan, Chokan Valikhanov, V.V. Radlov, N.A. Severtsov, N.M. Przhevalsky, V.V. Berthold etc. But the first practical steps to survey and beginning of scientific studies in the settlement were made in 1921 by archaeologist V.D. Gorodetsk, who continued work in 1924.

At the end of the 30s, archaeological excavations were carried out by an expedition under the leadership of academic A.N. Berenshtam, but they were discontinued due to the outbreak of World War II. In the postwar period, the excavation of ancient settlement was long and fruitfully carried out by I.I. Kopylov, A. Margulan, K.M. Baypakov, T.V. Savelieva etc. The result of all this painstaking work was the re-creation of history picture on ancient city (Tourism of Kazakhstan 2015).

Ancient Taraz settlement. It is located on the territory of modern central market of Taraz in the Zhambyl region.

Among the buildings of Taraz that have survived to our time, the mausoleums and mazars are of particular interest. Today they admire the architectural composition originality, decorative furniture and high building art. Fig. brickwork and new types of building material were used with great perfection in these buildings: alabaster, terracotta and tiled slabs of various shapes and ornaments.

Among 25,000 historical and cultural monuments currently in the Republic of Kazakhstan, about 1100 of them are located in the land of ancient Taraz. There are such architectural masterpieces of medieval architects as the mausoleums of Aisha Bibi, Babaji Khatun, Karakhan and Doutbek, as well as later unique monuments—the mosque of Abdykadyr, Abdul Ishan, Nametbai, Karakhoja, and mausoleum of Shokai Datkhi. All of them were restored and landscaped for visitors in framework of the program "Cultural Heritage".

Mausoleum of Karakhan (the mausoleum of Aulie-ata) was built over the grave on one of the representatives in the Karakhanid dynasty. Architectural monument of the XI century is located in Taraz city. It is a square in plan of portal-dome structure. It consists of central hall and three small corner rooms; the fourth corner of mausoleum is occupied by staircase leading to the building roof. The front facade of the mausoleum is facing south, framed by a minaret at the edges (Fig. 3.12).

In 1906, the mausoleum of Karakhan was rebuilt and, while preserving the construction principle, lost the original architectural and decorative furniture. Inside the mausoleum preserved gravestone. In 1982, the Karakhan mausoleum was taken under state protection (Tourism of Kazakhstan 2015).

Fig. 3.12 Mausoleum of Karakhan (*Resource* Available at https://Kazakhstan.travel/photobank)

Mausoleum of Dauytbek. It is located on the rabad territory in Taraz settlement, east of the Karakhan mausoleum. Among the local population, it is known under the name of Shamansur mausoleum.

The mausoleum was erected over the Mongol governors grave called Ulug-Bilge-Ikbalkhan Dauytbek. This is a portal-dome structure, square in plan. The portal is low, smooth with two small turrets on sides. Squat dome rests on the arches and walls of the mausoleum covering inner niches. The portal is close to the dome. In the centre of mausoleum interior, there is a gravestone in form of stone stepped pyramid. It has an inscription in Arabic, indicating the date of his death—1262. The inscription was read in 1893 by V.V. Barthold and in 1939 by A.M. Belenitsky—"The deceased owned a sword and quill. He was a defender of the Koran and faithful people" (Tourism of Kazakhstan 2015) (Fig. 3.13).

Ancient Zhankent settlement. Monument of Aral Zhankent, it is the capital of Oguz (one of the Turkic people). In medieval Arabic and Persian written sources this city is characterized as one of the major economic and cultural centres in lower reaches of the Syrdarya. The name of medieval Oguz capital, Zhankent (Yangikent) is translated as "new city", or "new capital". The medieval settlement Zhankent is located in the Kazaly district of Kyzylorda region.

City remains are fairly well-preserved until now. Settlement is almost rectangular, measuring 325 × 200 m. In the northwest part of settlement there is a well-preserved citadel. On the inner side of northern wall there is a separate manor, towering under the square shape. From the outside, a building structure was added

Fig. 3.13 Mausoleum of Dauytbek (own authors)

to the northern wall of settlement, which was preserved in the form of fallen wall remnants. Since 2007, the monument has been investigated within the framework of state program "Cultural Heritage". On the settlement citadel, several premises were investigated, which are part of large stationary dwelling.

In the interior and layout of this dwelling, the traditions of housebuilding inherent in the medieval urban culture of South Kazakhstan region and Khorezm are clearly visible. There are rooms for household and residential purposes. In residential areas along the walls were located "sufi". The premises were heated with small hearths. On the settlement citadel, a sanctuary was discovered, previously unknown structural features were revealed in construction of the fortress wall, and terracotta statue of two-headed ram was found, which is richly decorated with ornaments. A large collection of finds was presented by copper coins, tools, bronze products in the form of earrings, rings, bells, plaque patches that find analogies in the monuments of the end from the 1st—beginning of the 2nd millennium AD. Ceramics are represented by remnants of dining and household utensils in the form of hums, jugs, mugs, drinking bowls, tableware, made using the technique of hand-made molding and on a potter's wheel. The ceramic complex shows cultural traditions characteristic of the settled agricultural and nomadic tribes of South Kazakhstan region and Saryarka. The combination of various cultural traditions in the Zhankent pottery is probably one of the manifestations on close contacts in Turkic tribes in the framework of Oguz ethno-political association, which included various economic and cultural areas, and confirmed by written sources (Tourism of Kazakhstan 2015).

South Kazakhstan region is well-known for the holy city of Muslims–Turkestan, called the little Mecca, which has a number of outstanding religious shrines, which turned it into the largest pilgrimage centre. In the XII century, mausoleum of Ahmed Yasawi was built on the settlement territory. Nearby is the Aulie Kumshyk-ata mosque, crypts of the Kazakh statehood founders—Tauke, Tavakel, Dzhangir, Yesim, and Ablay mausoleums (Fig. 3.14).

In the very centre of Turkestan are well-known semi-underground "Eastern baths". On the early history of city (V–XII centuries) written sources are silent. The site of ancient Turkestan is an archaeological monument.

Turkestan is the most widely known among South Kazakhstan region cities. It was, according to written sources, in one day's transition from Otyrar on Syrdarya. Shavgar district is a modern Turkestan, and city of the same name is identified with the Shoy-tobe settlement, located 8 km south of Turkestan city. In the XII–XIII centuries Shavgar lost its trade and cultural significance. The city was abandoned and district centre was moved to Iassi-Turkestan. It, like many medieval cities, has

Fig. 3.14 The mausoleum of Ahmed Yasawi (own authors)

Fig. 3.15 Ancient Sauran city (*Resource* Available at https://Kazakhstan.travel/photobank)

flourished and declined. Iassi-Turkestan is located on a favourable trade route, and it became the residence of the Kazakh khans

Ancient Sauran city. Sauran is a medieval settlement in the south of Kazakhstan located in territory subordinated to the city of Turkestan, Turkestan region, 43 km north-west of the city, 1.5 km from the administrative border with Kyzylorda region, 2 km south of Samara-Tashkent highway, 500 m from railway (Tourism of Kazakhstan 2015) (Fig. 3.15).

Monument is part of state museum-reserve "Azret-sultan" with the centre in Turkestan. Modern village Sauran is located 14 km south-east of the settlement. In the VI–XIIIth centuries, city of Sauran was located 3 km south-east of the later city —now this ancient settlement is called Karatobe. The city on site of modern settlement Sauran functioned in the XIII–XVIII centuries, after which it was abandoned. Sauran was an important trade and economic centre that controlled the passage between the Karatau ridge and Syrdarya, along which the caravan route passed. Population of the city during its heyday is estimated from 3 to 15 thousand inhabitants. The settlement is oval in plan, extending from the north-east to south-west at 800 m, from the north-west to south-east at 550 m. Height of the fortress walls made of mud brick reached 7 m, the base of walls is 3 m, the width of gate is 5–6 m (Tourism of Kazakhstan 2015).

Sauran is place of the first discovery on Kazakhstan territory concerning kyariz system—an irrigation system of wells with underground canals for irrigating fields; in 2002, similar structures were found in three more places in the Turkestan oasis.

Arab geographer Makdisi wrote: "Sauran (Savran, Sabran)—a large city surrounded by 7 walls, cathedral mosque is located in the inner city". In the first half of the XIVth century, Sauran was the capital of the White horde. Here in 1320 the ruler of the White horde Sasy-Buk was buried. In 1514–1515, the poet Vasifi lived in Sauran, leaving a description of the city and its environs (Tourism of Kazakhstan 2015).

Otyrar settlement. Settlement represents the remains of well-known medieval Otyrar city, which is mentioned in almost all the geographical and historical writings of medieval cities in Semirechye and in registry of the Ispidzhab cities. Three entrances led to the city. Topography of Otyrar was formed mainly in the late Middle Ages (XV–XVIII centuries), so the citadel is not currently distinguished and can only be distinguished presumably. Excavation of Otyrar layers in VII–VIII centuries testify that it was an already established city. Remains of building structures made of mud bricks were found, a collection of dishes was collected, which was characterized by narrow-necked jugs with plums, lamps on column-shaped legs, coins of local coinage. The period of IX–XII centuries was time of the city life rise in south of Kazakhstan. New cities arise, old ones develop. Otyrar at this time occupied an area of 200 ha. (Tourism of Kazakhstan 2015).

Territory of the city was closely built up with houses grouped into quarter-blocks. In the IX–XII centuries public buildings like bathhouses appear in urban areas. In Otyrar, two baths were investigated on the rabad territory. Assembled collection of ceramics from IX–XII centuries and interesting materials as pottery kilns, also found in rabad. Rich collections of glass items have been collected such as fragments of decanters, bowls, glasses, bottles. Also there is lots of jewellery, such as beads from garnet, carnelian, glass, rings, and bracelets from bronze. The city flowering was interrupted by the Mongol invasion. In the middle of XIII century, Otyrar again turns into a large shopping centre (Fig. 3.16).

In the second half of the XIV century, South Kazakhstan region entered the state of Timur. Timur died in one of the Otyrar palaces. In Otyrar, Timur's name is associated with the construction of mausoleum over the grave of Ahmed Yasawi's teacher Arystan baba. The excavated mosque of Otyrar also belongs to the Timur period. The idea of urban development, nature of the city block, public buildings, formation of new types on Middle Ages housing became known from the Otyrar excavations.

Kazygurt. The legendary place (40 km from Shymkent) is located in the valley of the Talas Alatau. They believe that this is where Noah's ark landed. Locals call this place "Keme Kalgan" (Noah's ark place) (Fig. 3.17).

Mountain has a second landmark, where another mystical thing was found such as versatile cross made of stones. There are several versions of this unusual "sign" origin. Some researchers attribute this to Zarathustra. A huge cross may have been made with the ritual purpose of Zoroastrians, the followers of Zarathustra's training.

According to other versions, similar crosses—signs sent by Tengri, the god of heaven were made several centuries before appearance of Christianity in Altai, which was the ancient homeland of the Turks.

Fig. 3.16 Otyrar settlement (own authors)

Fig. 3.17 Kazygurt (authors)

Shilter-ata is a the natural complex, also included in the list of holy places. The Ata-ana rock symbolizes the creators of human race, Adam and Eve. Two stones with a height of 33 m have a passage between them, through which only a pious person can pass. In the mountains of Kazygurt and in surrounding area, 33 archaeological sites were found, most of which are located in upper mouth of the Keles river and its tributaries (Tourism of Kazakhstan 2015).

The Great Silk Road passed through Kazygurt. From here, there is a start of the shortest routes to Uzbekistan, ancient Tashkent, Bukhara, Khiva and Samarkand.

Mausoleum of Abdel-Aziz-Bab, XV–XIX century. According to historical data, in 766 a detachment of Arab volunteers led by Iskak-baba stepped onto the territory of modern Kazakhstan. The standard-bearer of this detachment was Abdel-Aziz-bab, who in the spread of Islam showed real heroism. His erudition, resilience and unshakable loyalty to the cause of Islam were immediately apparent. The saint, who knows how to discourage misfortune and cleanse from slander, was nicknamed by the people—Balegerdan (expels trouble). Under this name, he remained in the history (Fig. 3.18).

Petroglyphs of the archaeological Tamgaly landscape. Tamgaly tract is located in the Anrakai mountains. Most of the petroglyphs are located in lower part of the main gorge and in the side gorge adjacent to it from west, as well as in 7 sais located northwest of main gorge (Fig. 3.19).

Total number of drawings in main gorge is about 2000. All of them are conventionally divided into 7 groups. There are images of sun-gods, mummers, warriors-soldiers, marriage couples, women in labour, as well as multi-Fig. compositions depicting people and animals, scenes of hunting for animals and scenes of

Fig. 3.18 Mausoleum of Abdel-Aziz-Bab (own authors)

Fig. 3.19 Petroglyphs of the archaeological Tamgaly landscape (*Resource* Available at https://Kazakhstan.travel/photobank)

bull sacrifice. The plots depicting chariots are rare. There are numerous solar signs. Most of the petroglyphs belong to the Bronze Age (Tourism of Kazakhstan 2015).

Drawings, made in the Sak "animal" style, are placed mostly separately from the more ancient petroglyphs, but in some cases complement or even overlap them. Medieval rock paintings are carved on hills surrounding the gorge and in the adjacent waterless sais. The central complex stands out for the densest concentration of petroglyphs and, presumably, altars, which indicates the possible ceremonial purpose of this place for sacrifices. In addition to petroglyphs in Tamgaly, a large number of ancient tombs were discovered: middle and late Bronze Age stone boxes-cysts, mounds of earth and stones from the early Iron Age.

Kurgan burial ground Besshatyr. It is located on the territory of state national park "Altyn-Emel", 170 km from Almaty.

The funerary chambers were under mounds of stone and gravel, almost well preserved, only two mounds of the chamber were burned. Rich burial was uncovered in the Tenlik mound, among the items of decoration on clothing made of sheet gold, there was also an iron rod encrusted with gold. On some vertically standing fencing plates there are embossed tamgas of the Kazakh clans. Five kilometres to the north-east of large barrows 7 fences are located, similar in shape and design to the fences of Besshatyr barrows, and 10 km to the west of the royal borders a chain of 45 stone plates with animal carvings is from north to south.

These structures mark the boundaries of ritual site on Besshatyr necropolis (Tourism of Kazakhstan 2015).

Pyramids of Besshatyr architecturally complete the surrounding space, giving it timeless charm and admiration for enigmatic mystery of these places.

3.2 Ecological and Tourist Infrastructure

One of the basic concepts of tourism geography is the notion of "tourist attractiveness" on locality or object. Attractiveness of particular area is determined not only by the presence of favourable the natural conditions or historical sites, but also by presence of developed tourist economy—the tourism infrastructure.

Tourism industry is a certain complex of various sectors, the creation and development of which is stimulated by arrival and stay of tourists. This economy consists of mutually complementary industries that can satisfy the needs of tourists only as a complex (Yerdavletov 2015).

Tourism industry is divided into two groups: direct and indirect.

- direct—these are institutions and objects intended exclusively for serving tourists: tourist bases, hotels, sanatoriums, hostels, transport centres, etc.
- indirect—institutions and objects of non-tourist destination, but used partly or periodically by tourists.

There are various tourist facilities on the territory of Kazakhstan.

According to the generally accepted structure of tourist economy, institutions and facilities designed to serve tourists are divided into four bases:

(1) overnight stay (accommodation);
(2) food and beverage;
(3) transportation;
(4) accompanying.

Base of overnight stay and accommodation for tourists is considered as the most important element of tourism. Otherwise, it is called the hotel industry. Hotel industry is the essence of hospitality system, which proceeds from ancient traditions in the history of mankind—guest respect, solemnity of its reception and service.

Today, the hospitality industry of Kazakhstan is a fairly developed system of regional economy.

In accordance with the recommendations of UNWTO, all accommodations in the Republic of Kazakhstan are represented in two categories: collective and individual. Collective tourist accommodation facilities are divided into:

(1) hotels and similar accommodation facilities for tourists (hotels, similar establishments);
(2) specialized institutions (recreational facilities);

(3) other collective institutions;
(4) individual accommodation facilities.

Hotels. From the main signs of hotels it should be noted, first of all, the presence of room numbers at least 10 units. Hotels provide a list of required services: room cleaning, daily bed making and sanitary facilities, as well as an extensive range of additional services. Depending on the specific equipment and features of provided services, hotels include wide-ranging hotels, flat-type hotels, motels, roadside hotels, resort hotels, clubs with accommodation, and others.

The group of establishments similar to hotels, has a number of rooms and provides a list of mandatory services. These are the guest houses, furnished rooms, tourist hostels and so on and so forth.

Specialized institutions (with unified management) are also intended to serve tourists. They do not have rooms. Here the initial unit can be a dwelling, collective bedroom, playground. In addition to providing accommodation for tourists, the institution can carry out other activities. Examples of such specialized institutions are health facilities (health centres, resorts, sanatoriums, health farms, etc.), camps, accommodation facilities in public modes of transport (trains, ships), and in congress centres.

Other collective accommodation facilities include, first of all, dwellings intended for recreation. Examples of these accommodation group facilities are flat hotels, house complexes or bungalows. These premises have unified management, are rented for a fee or rent, free of charge by any person (people) or organization. The structure of other collective accommodation facilities includes objects (with a single management) at camp grounds, in bays for small vessels. The client is provided with an overnight stay and number of services such as information, shopping, and leisure. Tourist hostels, youth hostels, school and student dormitories, nursing homes and similar social facilities should also be considered as other collective accommodation facilities.

Individual accommodation facilities are provided for a fee, rent, free of charge. This includes dwellings (flats, cottages, mansions), which are rented alternately by members of the household (time-sharing). Tourist can also stay in a family house on loan. Tourist can be rented from a private person or agency in fully equipped dwelling (house, flat, cottage) or stay for free with relatives and acquaintances.

Main means of accommodation in tourism industry of the Republic of Kazakhstan is a hotel. Hotels in Kazakhstan differ in the number of permanent beds (capacity). In terms of comfort, in accordance with international practice, hotels in Kazakhstan differ in stars: from the lowest (1 star) to the highest (5 stars). Differentiation of hotels by star occurs in accordance with the regulatory requirements for hotels on various categories (Table 3.1) (Tourism of Kazakhstan. Statistical compilation 2017).

In recent years, Kazakhstan's rapid economic development has consistently experienced a shortage of hotels in all classes, especially in two financial and administrative centres of republic as Nur-Sultan and Almaty. Today, the Kazakhstan hotel market is mainly represented by three, four star hotels.

Table 3.1 Distribution of placements by category

Hotels, years	2012 year	2013 year	2014 year	2015 year	2016 year
5-star	16	17	20	20	20
4-star	51	53	61	75	85
3-star	98	103	122	138	138
2-star	27	22	20	19	20
1-star	24	17	19	17	15
No categories	875	955	1184	1256	1441
Other placements	435	511	630	813	1035
Total	1526	1678	2056	2338	2754

The largest five-star hotels in Kazakhstan are "Rakhat Palace", "Rixos Almaty", "InterContinental Almaty", "Donatello Boutique Hotel", "Royal Tulip Almaty", four-star hotels, for example, "Grand Hotel Eurasia", "Otyrar", "Dostyk", "Astana Hotel", "Best Western Plus Atakent Park Hotel", "Kazzhol", "Worldhotel Saltanat Almaty", "Holiday Inn Almaty", "Ambassador" and three-star hotels "Kazakhstan", "Alma-Ata", "City Hotel Tien Shan", "Zylikha", "Renion", and others are located mainly in the southern capital, Almaty city.

Network of modern hotels built in Nur-Sultan. To date, there are about 100 hotels in Nur-Sultan, most of which have 20 rooms. Main market of accommodation facilities in Nur-Sultan is represented by mini-hotels from 4 to 30 rooms. The largest share is 54%, concerning all existing objects in hotel real estate falls on objects of the 1 and 2 star class. More than 20 hotels in Nur-Sultan can be attributed to 3 stars (for example, "Abay", "Tengri", "Sary-Arka", "Europe Palace", "Turkestan", etc.). Their share in the total volume is 41%. Intermediate level between a luxury 5 star hotel and 3 star hotels is occupied by 4 star hotels, for example, "Comfort-Hotel Astana", "Grand Park Yesil", "Astana-Park Hotel", "Mukammal", "Business Hotel" and others. There are 4 large hotels with more than 150 rooms, 3 of which belong to the class of 5 stars. These are the "Rixos President Hotel Astana", "Okan InterContinental Astana", "Radisson SAS Arman".

Main statistical indicator is the material and technical base of placements. These include: number of rooms; one-time capacity; occupancy rate of hotels; number of provided bed-days; average cost of bed-days; scope of services provided by accommodation facilities; number of visitors served by accommodation facilities (Table 3.2) (Tourism of Kazakhstan. Statistical compilation 2017).

According to statistics, the highest one-time capacity in the Republic of Kazakhstan in 2016 was noted in Almaty, East Kazakhstan, Karaganda, Nur-Sultan and Akmola regions. Generally, in Kazakhstan there is an increase in one-time capacity.

Number of visitors served by accommodation sites in the whole of the Republic of Kazakhstan reflects a steady upward trend from 3.0 million people, in 2012 to 4.2 million people, in 2016, as well as a steady trend is observed in the number of locations from 1526 in 2012 to 2754 in 2016 y.

Table 3.2 Material and technical base of placements for 2016 year

Regions, indicators	Number of rooms, units	Nonrecurrent capacity, beds	Provided bed-days	Bed occupancy rate in percentage	Number of serviced rooms	Average cost of a bed-day, tenge
Republic of Kazakhstan	60,427	138,062	7,582,785	24.0	4,261,826	4979
Akmola	4590	11,490	546,460	17.4	233,250	3906
Aktobe	1789	4382	260,993	24.3	118,486	5096
Almaty	6457	16,091	655,987	22.6	280,490	2965
Atyrau	2469	3876	351,073	27.8	299,304	8093
West Kazakhstan region	1546	2878	304,244	29.5	179,907	5754
Zhambyl	1776	3172	160,562	15.6	101,527	5257
Karaganda	4637	12,441	579,701	23.1	294,398	5233
Kostanay	1817	5551	274,429	23.3	149,842	3934
Kyzylorda	1020	1789	105,568	18.8	63,334	6126
Mangystau	3035	5462	287,843	21.9	228,627	7165
South Kazakhstan region	3116	7634	374,328	21.5	180,083	5248
Pavlodar	2900	7212	251,559	19.3	96,875	4524
North Kazakhstan region	1357	3870	177,477	21.7	83,355	3746
East Kazakhstan region	8320	25,847	996,129	27.2	438,479	1876
Nur-Sultan city	7002	12,183	1,023,830	27.6	713,759	9886
Almaty city	8596	14,184	1,232,602	28.2	800,110	12,023

Along with hotels and similar hotel-type establishments, ecotourists make extensive use of specialized and other collective accommodation facilities. Among such accommodation means Khrabovchenko V.V. includes tourist shelters, alpine camps, tourist camps, fixed campgrounds, base camps, overnight stays, etc. (Khrabovchenko 2004).

Specialized accommodation facilities differ from hotels by the minimum set of mandatory and additional services provided, the small comfort of rooms and cottages, and also the maintenance of mainly certain categories of tourists (fishermen, hunters, climbers, etc.).

Tourist bases are hotel-type enterprises that provide accommodation, meals, and tourist-excursion, cultural, social, and physical-fitness services to groups of tourists.

Climbing camp is a sports base intended for training climbers and strengthening their skills.

Mountain huts and shelters—belong to various tourist, climbing clubs, federations for the short-term stay of climbers before and after climbing, and while waiting for bad weather.

Parking—the location of tourist group for night or rest. There are several types: bivouac, overnight stays, base and intermediate camps. Tents and sleeping bags are used at parking.

Ecotourists are also widely use the holiday homes, recreation centres, chalets, mountain hotels, etc.

Travellers with a view to ecotourism increasingly prefer accommodation facilities located in the natural environment, away from the noisy city bustle.

Food and beverage base is an integral part of tourism industry. In every village there are canteens, restaurants, cafes, bars and other catering organizations (Table 3.3) (Tourism of Kazakhstan. Statistical compilation 2017).

Currently, a large number of catering enterprises are opening up in the catering market of Kazakhstan and the volume of food and beverage services is growing at significant pace (Table 3.4) (Tourism of Kazakhstan. Statistical compilation 2017).

Steady increase in the volume of catering services in Kazakhstan can be traced from 2012 to 2016. So, if in 2012 the volume of services in enterprises amounted to 220,484.2 million tenge, then in 2016 its volume reached 415,836.9 million tenge, i.e. growth 1.89 times.

Table 3.3 Number of restaurants, cafes, bars and canteens

Food enterprises, years	2012	2013	2014	2015	2016
Restaurants	1831	1910	1765	2008	2272
Cafes, bars	9117	10,098	12,099	9254	8366
Canteens	6516	8730	9516	8210	8478
Other catering facilities	2744	2503	3335	3397	2318
Number of seats, units	1,047,085	1,196,199	1,358,232	1,287,269	1,334,857

Table 3.4 Volume of food and beverage services by region (million tenge)

Regions, year	2012	2013	2014	2015	2016
Republic of Kazakhstan	220,484.2	238,307.7	269 320.5	321 919.0	415 836.9
Akmola	3987.7	4723.5	7 203.4	8 392.8	9 729.7
Aktobe	9648.7	9907.0	11 051.4	12 619.4	16 044.3
Almaty	9227.8	9997.0	11,410.7	11,473.2	11,024.6
Atyrau	21,154.8	21,048.3	22,951.1	26,107.1	41,808.0
West Kazakhstan region	15,155.4	11,345.3	10,734.4	15,315.2	20,464.8
Zhambyl	4777.1	5818.9	6621.4	7176.1	8357.4
Karaganda	11,614.4	13,001.9	14,578.1	16,604.5	17,240.8
Kostanay	4752.2	5130.2	5585.1	6200.0	7669.7
Kyzylorda	8107.1	11,934.9	12,932.4	11,253.0	10,813.8
Mangystau	18,913.4	18,843.8	23,104.3	22,696.9	22,887.1
South Kazakhstan region	14,242.4	18,671.4	22,565.9	26,488.3	28,785.6
Pavlodar	8545.3	7202.8	8618.9	12,090.2	9623.5
North Kazakhstan region	2712.9	3905.9	4487.7	3571.9	4110.0
East Kazakhstan region	16,861.8	19,652.8	19,808.1	19,576.7	19,176.2
Nur-Sultan city	20,721.3	24,397.9	30,673.4	41,847.5	56,724.1
Almaty city	50,061.9	52,725.3	56,994.1	80,506.2	131,377.3

It should be noted that the organization of catering have to pay great attention on national cuisine development. After all, exotic dishes and an interior decorated in national style are also fascinating to tourists.

National cuisine is one of the important factors in attracting foreign tourists to the country. This is understandable because in this kitchen national peculiarities and traditions of the people are firmly preserved, which are always interesting for representative of another country. Tourists who come to Kazakhstan, of course, are willing to try traditional local dishes. The basis of the Kazakh national cuisine has long been the livestock products—meat and milk. Later, due to increased contacts with sedentary people and agriculture development, the Kazakh began to use flour products. Now vegetables and fruits have become common on their table The Kazakh gastronomy currently has all the signs of globalization.

In the summer, almost in every Kazakh family (we are talking, of course, about the countryside inhabitants) prepare two old national drinks, which have long been called "heroic drinks". Name of these drinks is koumiss and shubat. The first of them is fermented mare's milk in a special way, known for its healing properties far beyond the borders of Kazakhstan. The second is obtained similarly from camel milk. Also in Kazakh families, another drink is also widely prepared—airan-sour milk. It is drunk as a soft drink, they are filled with various cereal broths. They make kurt and irimshik from airan. A kurt is salted cottage cheese rolled into balls and dried in the sun, and irimshik-fat crumbly cottage cheese (Yerdavletov and Uvarova 2001).

Favourite drinks of the Kazakh, except for koumiss and shubat, include strong black tea, which they drink at any time of the year. Usually milk or cream is added to the tea.

In winter, the Kazakh are dominated by meat food. The most favourite and common meat dish is besbarmak — meat and pieces of rolled dough cooked in broth. Besbarmak in Kazakh means “five fingers”, because this dish is usually eaten by hands from large plate, which is placed in the middle of table. Meat in besbarmak served with bones. In this case, the owners immediately chop it, and bones with meat are distributed to guests. Sheep head is presented to the guest of honour who cuts off the ears and gives them to young men, girls are given the tongue. After that, the distinguished guest cuts off small pieces of meat from the head and distribute them to everyone sitting at the table. Then the head moves from one guest to another. Besbarmak with sorpa is a strong meat broth, in which meat and dough are cooked. The broth is served in large bowls.

Usually, when eating, besbarmak is preceded by kuyrdak, a roast of liver, lung, and meat with potatoes. As a cold appetizer, liver is sliced into thin pieces, with boiled sheep-fat (kuyryk bauyr), and also tasty raw smoked sausages from horse meat—kazy and shuzhuk. The integral attribute of the Kazakh table (dastarkhan) is almost always the baursaks,—pieces of sour dough, fried in oil. Kazakh also bake cakes from sour and unleavened dough, often eating tare (toasted millet) with milk or tea.

In general, the Kazakh, being a nomadic people, have long been able to harvest meat for the future. They dry, smoke and salt it. In this regard, such kind of Kazakh food as kul azyk is of particular interest. If guests of Kazakhstan can get acquainted with the national dishes listed before in stationary conditions, then with kul azik—only in the conditions of hiking. The fact is that this is camp food, and it is perfectly suited for tourist trips and travels. What is kul azyk? A sufficiently detailed idea of this already largely forgotten food can be derived from an analysis of ethnographic sources (Yerdavletov and Uvarova 2001).

Previously, this dish is similar to the oatmeal usually used by warriors going on a hike. It was enough to dissolve one tablespoon of it in boiled water and drink it in order to be full. The method of cooking kul azik is as follows: slaughtering a fat sheep (horse, cow, yak), finely crumbled meat and fat separately, all fried and dried for a long time in the sun. At the same time, the dough was prepared and on the rest part of fat they baked salt cakes, which were also dried in the sun and milled on the millstones. The ground meat and tortillas were mixed.

Thus, the kul azyk was intended mainly for a long journey. And therefore, it can be of double interest for tourism: as an exotic dish and as hiking food. How can kul azyk be introduced into the tourist service? It seems that in the future mobile tourist camps will be organized in the mountainous regions of Kazakhstan, where equestrian routes will be arranged. This is where it can come in handy because kul azyk does not require refrigeration and is a high-calorie product. In addition, it takes up little space, which is important in the field of conditions. Kul azyk would be very convenient food for horse and hikers, going to hike for 1–2 days.

In the Republic of Kazakhstan there are various forms of service:

- "a la carte" service, i.e. free choice by the client of dishes from the menu offered by the restaurant. It is usually used when servicing individual tourists;
- "tabledot service", i.e. according to a single menu for all customers without the right to choose dishes. In most cases, it is used in the organization of catering for tourist groups;
- "buffet", i.e. free choice of dishes exhibited at the common table and use of the self-service form.

Restaurant market in Kazakhstan is also divided into several segments: fine dining, casual dining—up to 80% of the market; fast-service enterprises (quick service restaurants), represented by network companies, local institutions, as well as Turkish substitutes of world brands without copyright and the original name; canteens and cafeterias (fast casual).

All of them are widely in demand among ecotourists.

Transport base. Transport plays an important role in organizing the infrastructural support of the tourism and recreation process.

Transport is an integral part of tourism industry infrastructure and is one of the main services provided to travellers. Quality of transportation in the organization of transport trips largely depends on normal functioning of the transport park and infrastructure.

Transportation is an integral part of tourist product formation, the most important component of material and technical base of tourist industry. According to the UNWTO classification, transportations are divided into the following groups (Tourism of Kazakhstan 2015):

- air transport, including: scheduled flights, charter flights, passenger transportation by small aircraft;
- water transport, including: passenger lines and ferries, cruises, and other transportation;
- land transport, including: railway transport, intercity and city buses and other public road transport, private cars with capacity of up to 8 people, rental vehicles, and other land vehicles.

In modern tourism involved 4 main types of transportation: road (bus), railway, air, water.

Transportation system of Kazakhstan plays an important role because of territorial and geographical features of the republic, specific of which are: the vastness of geographical space and diversity of the relief important for construction of transcontinental road communications, relatively low population density, contrast of geographical and climatic conditions, and remoteness from major foreign markets and main economic centres of republics and, as a result, heterogeneous economic development territories.

Table 3.5 Operating length of public railway tracks (km)

Regions, year	2012	2013	2014	2015	2016
Republic of Kazakhstan	15,333	15,341	15,341	15,341	16,104
Akmola	1559	1559	1559	1559	1559
Aktobe	1443	1443	1444	1444	1499
Almaty	1394	1402	1402	1402	1401
Atyrau	742	742	742	742	742
West Kazakhstan region	431	431	431	431	431
Zhambyl	1103	1104	1104	1104	1104
Karaganda	1940	1940	1940	1940	2467
Kostanay	1271	1271	1271	1271	1336
Kyzylorda	755	755	755	755	871
Mangystau	926	926	926	926	926
South Kazakhstan region	552	552	552	552	552
Pavlodar	925	925	925	925	925
North Kazakhstan region	807	807	807	807	807
East Kazakhstan region	1209	1209	1209	1209	1209

Transport accessibility of tourist facilities is not so much an element of improvement, as the conditions for tourism development in the area. On the other hand, the growth of tourism contributes to expansion of road construction. Transport as the most important industry providing services of tourist type, has an integration character, as it unites individual regions into a single one. Therefore, the most urgent is the question of eliminating possible gaps in transport message.

Railway transport. The operating length of public railway tracks is 16.1 thousand km. (Table 3.5) (Transport in the Republic of Kazakhstan. Statistical compilation 2017).

Currently, JSC "National Company" "Kazakhstan Temir Zholy" is the only carrier of goods on market of the Republic of Kazakhstan. At the same time, for transportation of JSC "NC" "KTZh" uses the rolling stock of JSC "Lokomotiv", JSC "Kaztemirtrans", inventory cars of other railway administrations and private fleet of operator companies (Table 3.6) (Transport in the Republic of Kazakhstan. Statistical compilation 2017).

According to statistics, in 2016, more than 22 million passengers were transported. The high popularity of rail transport among tourists is due to relatively low fares, allowing for a lower fee to visit more cities. To the services of this transport type, which is resorted mainly to those tourists for whom the motive of saving time is not the leading one. Although the essence of rail transportation causes additional costs, but does not require special investments.

Aviation transport. Particularly there is a large role of tourist transportation by air transport, as it is fairly high-speed and comfortable. All aviation enterprises are interested in attracting as many passengers as possible. This is achieved by introducing various preferential tariffs in the period of easing passenger traffic. Usually,

Table 3.6 Transportation of goods and passengers by all modes of transport in 2016

Regions, indicators	Transported goods, baggage, cargo luggage, thousand tons	Freight turnover, million ton-kilometer	Transported passengers, number of people	Passenger turnover million passenger-kilometer
Republic of Kazakhstan	3,729,184.0	518,602.2	22,332,790.6	266,784.2
Akmola	113,105.5	4992.6	1,700,076.9	7278.0
Aktobe	76,322.7	6126.1	278,421.4	14,911.6
Almaty	178,961.3	8042.2	808,282.5	14,107.6
Atyrau	145,093.3	54,199.1	181,019.7	1393.0
West Kazakhstan region	39,414.9	2906.3	434,331.5	8409.7
Zhambyl	91,250.6	2735.1	957,286.5	7838.6
Karaganda	727,476.3	10,355.1	2,418,111.3	38,559.0
Kostanay	277,258.9	10,527.5	1,751,732.0	17,412.1
Kyzylorda	104,294.5	13,877.6	335,468.9	5991.7
Mangystau	218,917.0	8256.4	84,801.4	4322.9
South Kazakhstan region	158,747.2	9103.0	2,478,472.7	20,308.0
Pavlodar	131,462.3	32,742.2	1,244,492.8	25,153.3
North Kazakhstan region	48,181.2	3445.9	506,797.7	3917.7
East Kazakhstan region	582,434.4	15,236.5	1,705,150.7	22,187.4
Nur-Sultan city	142,553.6	10,570.7	2,441,056.8	18,916.2
Almaty city	252,265.6	23,745.7	4,984,214.7	38,163.5

these discounts are related to the condition of buying a return ticket, their number or depending on the season. Air transport also performs the functions of tourist enterprise: it becomes the base of an overnight stay for the time being, provides food and additional services. During the flight, it is possible to organize a kind of excursion, telling, for example, about the country of reception, etc. It is assumed that the role of air transport in tourist traffic will increase everywhere. First, it is the expansion and improvement of material and technical base, the improvement of technical and operational qualities of aircraft, as well as legal issues development related to border crossings (Table 3.7).

Table 3.7 Passengers transported by public air transport (thousand people) (Transport in the Republic of Kazakhstan. Statistical compilation 2017)

	2012	2013	2014	2015	2016
Republic of Kazakhstan	4512.0	4985.6	5434.6	5922.8	6021.7
Almaty	30.0	25.5	28.5	25.2	16.2
Atyrau	0.4	–	–	4.3	0.1
West Kazakhstan region	59.7	140.1	354.7	332.4	x
Karaganda	7.2	10.0	8.5	x	x
Mangystau	–	–	–	–	–
South Kazakhstan region	1095.8	1109.6	1254.7	1224.7	1288.9
Pavlodar	–	–	–	–	–
East Kazakhstan region	5.1	–	–	–	–
Nur-Sultan city	5.5	1.3	1.4	5.2	34.6
Almaty city	3308.3	3699.1	3786.8	4324.4	3945.5

There are 50 airlines and aircraft operators are operating in the republic. In domestic flights, there are regular flights on 46 routes. The Number of passengers transported by airlines of the Republic of Kazakhstan increased from 4.6 million people in 2012 to 6.0 million people in 2016.

Road transport. Road transport is of great importance in expanding domestic and foreign trade, the formation and strengthening of interstate connections. The Feature of road transport is mobility, accessibility, manoeuvrability. For many destinations in the regions of the country, this is the only way to transport goods and deliver passengers.

Road transport in recent years has become a demand for the transport of tourists because highways have access to Russian cities and further to Europe, neighbouring states: Kyrgyzstan, Uzbekistan, and China. Some tourist enterprises in North and West Kazakhstan have already mastered low-priced auto-tours to Germany and Poland.

Water transport. Currently, the most popular transport in the international tourist market is sea and river. This type of transport in Kazakhstan is underdeveloped, which is explained by the less developed hydrographic network. Currently, coastal transport infrastructure of Kazakhstan on the Caspian Sea includes two active seaports—Aktau and Bautino.

So, transport infrastructure is the most important condition for tourism development. Currently, requirements for technical safety of vehicles, speed, comfort, etc. are strengthening. In the end, these factors that will be decisive in choosing one or another means of transportation, and transport services will be expanded depending on specific features of various transport types concerning tourist requirements.

Accompanying base. In addition to transportation, accommodation and catering, among the tourist services there are also services, the so-called accompanying base.

Associated enterprises include a set of institutions involved in providing tourist and recreational services with necessary goods, tourist and sports equipment, supplying raw materials, fuel, semi-finished products and other types of products necessary for the technological process in tourist industry.

These include sports, alpine skiing, alpinist complexes, horse farms, hippodrome, hunting and fishing farms, subsidiary farms (for the production of eco-friendly products), beach rest places, and first aid stations.

The largest sport complex in Almaty region is Medeo. It is the world's highest mountainous sport complex, near the Small Almaty river, and equipped with the world-well-known high-altitude skating rink with stands for 10 000 spectators.

The Almaty republican hippodrome, founded in March 1930, is the venue for not only testing riding and trotting horses, but also various equestrian events, championships, competitions of republican scale.

Viewing platforms of tourists who want to explore all the surroundings in one fell swoop. The most well-known observation place is the Koktobe, which is located in Almaty. To look at the city from a bird's-eye view … everyone has a similar look that causes their emotions. After all, the view that you will discover is probably the best in the city! You can watch with the naked eye, and you can do it with help of special binoculars. It is so wonderful to look down and realize that the city is a huge being that way, and will remain forever in the heart. It is especially wonderful at night when the rivers of lights pulsate with light, sparks of car headlights are flowing in an endless stream, and you look at this beauty from a height and enjoy. Urban landscapes have their own special appeal. Take a look! Rise above the bustle and once again make sure of the Almaty beauty. To achieve this, you only need to climb to the observation deck.

Hunting and fishing farms offer ecotourism services such as photohunting and birdwatching (in the wild), and sport fishing. So, in 2009 the World Championship in Sport fishing was held on the lake Balkhash for the first time in Kazakhstan.

References

Bayandinova S, Mamutov Zh, Issanova G (2018) Man-made ecology of east Kazakhstan. Springer, Singapore, p 144

Bayandinova S, Mamutov Z, Issanova G (2018) The natural factors of formation and development of geosystems in east Kazakhstan (book chapter). In: Environmental science and engineering (subseries: environmental science), pp 1–30

Khrabovchenko VV (2004) Ekologicheskiy turizm (Ecological tourism). Finance and Statistics, Moscow, p 208

Transport v Respublike Kazakhstan. 2012–2016 (2017) Statisticheskiy sbornik (Transport in the Republic of Kazakhstan. Statistical compilation). Astana, p 63

Turizm Kazakhstana (Tourism of Kazakhstan) (2015) Bastau, Almaty, p 520

Turizm Kazakhstana 2012–2016 (2017) Statisticheskiy sbornik (Tourism of Kazakhstan. Statistical compilation). Astana, p 64

Yerdavletov SR (2010) Geografiya turizma (Tourism geography). Kazakh University, Almaty, p 412

Yerdavletov SR (2015) Geografiya turizma Kazakhstana (Geography of tourism in Kazakhstan). Kazakh University, Almaty, p 180

Yerdavletov SR, Uvarova AK (2001) Natsionalnaya kukhnya kazakhov kak stimul razvitiya inostrannogo turizma v Kazakhstane/Vozmozhnosti razvitiya turizma Sibirskogo regiona i sopredel'nykh territoriy (National cuisine of the Kazakh as an incentive for foreign tourism development in Kazakhstan/Opportunities for tourism development in the Siberian region and adjacent territories). Tomsk, pp 88–92

Potential Geoparks of Kazakhstan

4

Kazakhstan is rich in all kinds of unique geological and geomorphological objects, which can be the basis for development of geological tourism, that is, they can become the basis for the formation of national geotouristic market and creation of a geological parks network.

The natural attractions formed in various geological eras are located in different parts of this huge country. Currently, various monuments of geological heritage are under state protection within national parks, reserves and in territories where active economic activity is carried out. These include unique landforms, sources of mineral thermal waters, healing mud, monuments of palaeontology and areas of relict plants. In the list of the natural monuments of republican significance there are 26 the natural complexes, valuable in ecological, scientific, cultural and aesthetic terms. They are assigned to the objects of the state nature reserve fund.

Currently, the most interesting and accessible places are included in ecological and excursion routes, sanatorium and resort activities have been created on mineral water deposits and medical mud.

Kazakhstan, which has unique relict resources, the creation of geoparks is a topical and timely issue. The formation of geoparks will contribute to the development of geotourism and will attract many tourists from all over the world, since the country has unique forms and types of relief, geological monuments that can reveal many secrets of the Earth development, as well as attract the attention of the international community on their research.

The natural unique landscapes are resources that can be the basis or base of geoparks that will help develop scientific and applied research in the field of Earth sciences, and this, in turn, will contribute to the development and strengthening of geotourism.

Kazakhstan has all the conditions for the development of geo- and eco-tours in unique the natural and cultural-historical places, which will increase the demand for national tourist product.

In recent decades, Russia, Portugal, Greece and other states have developed geotourism and have a great interest in expanding its borders. This type of tourism

K. Iskakova et al., *Ecological Tourism in the Republic of Kazakhstan*,
https://doi.org/10.1007/978-3-030-77462-2_4

contributes to the development of regions and increases the scientific potential and well-being of the country.

Geoparks will be able to ensure the preservation of geological monuments of the Earth's history and transfer the inheritance for future generation, which are found in almost every region of Kazakhstan. To achieve this, you should select significant objects of geological heritage, based on the history of the Earth's development, compile a catalogue with a potential list of unique nature objects to facilitate their search, and at best a register of such objects. In general, in sovereign Kazakhstan there is a need to create a single state register of tourist resources. In Kazakhstan, it could be called the «National Register of Tourist Resources of Kazakhstan».

Development of this document will contribute to the assessment of tourism and recreational the natural resources for its development. Implementation of such work will make it possible to have a significant positive impact on the development of domestic and inbound tourism, as the most economically attractive types.

4.1 Unique Relief Shapes

In Kazakhstan, there are all forms and types of Earth relief. Out of these, the oldest form of relief is peneplene, which is located on the territory of Kazakhstan—this is the denudation platform lowland of Saryarka. At the end of the XIXth century, this term was used by the founder of geomorphology, William Morris Davis.

He developed and introduced the term «peneplene» (almost a plain) in the XIXth century. At the beginning of the XXth century, in 1903, a scientist made his only trip to Central Asia and Kazakhstan.

Peneplenes are classical parts of the Earth's surface relief. In North America—these are the ancient gentle mountains of Appalachian, in Eurasia—the old relief of widely spread denudational plain of Saryarka, areas which are presented as folded-block low mountains of Kokshetau, Ulytau, Bayanaul, Karkaraly, Bektau-ata, Kyzylarai, Bugyly. All of them are preserved against the background of denudation plains—the transformed peneplated plateau (Woodward et al. 2018).

For many millions of years, Saryarka in Central Kazakhstan, which was a mountainous territory, has turned into a gentle plain, which is now called the Kazakh shallow settlement.

Classical pre-Paleozoic formations are transformed into folds and greatly changed. Often they are permeated by eruptive rocks—the natural granites and are associated with elevated areas of the relief.

Dominant comparative uplands of the Kazakh placative country are about tens of meters. However, at times between the hills, low ridges and ridges there are peaks that clearly stand out against the background of encircling terrain. These are, for example, the Kokshetau and Bayanaul mountains. Kokshetau mountains are located 60 km southeast of Kokshetau. They are an expressive category of granite unusually weathered hills with pine forests. Their length is around 20 km, the maximum point is the Sinyukha hill—reaching 887 m above sea level (Fig. 4.1).

Fig. 4.1 Kokshetau region (authors)

Near the Kokshetau hills, the beautiful fresh lakes of Borovoe resort area are located closely, where we see sanatoriums, holiday homes, and tourist bases.

Mostly known resort Borovoe, deployed for saving the same lake. The south-western Bayanaul mountains of Pavlodar region are beautiful and attractive. This is a complex mountain-bearing massif, including various groups of mountain hills of various shapes. Their highest point—Akbet city—with a height of 1027 m. Granite hills with pine forest are located nearby of large basins, where the fresh lakes Sabyndykol and Zhasybay are located. On the shores, where holiday homes and pioneer camps are located.

The northern part of Kazakhstan is characterized by reduced areas, which include river valleys and lakes with an equalized relief. Thus, located in the Akmola region, the Tengiz-Kurgalzhyn group lies at an altitude of 300 m above sea level. Low mountains in the west gradually close to plains of the Turgai plateau, and from the north to plains of the West Siberian plateau.

The Turgai plateau is located almost throughout the Kostanay region, except for the northern part. Basically, the plateau consists of almost horizontally lying monotonous, porous clay-sand sediments of tertiary age. The Rail of the Turgai plateau is characterized as disjointed, sections with heavily divided ravines, with sections of the plateau concerning group and single remains, as well as wide river valleys. Remainers, or «turtkuli», are a group or single boulders with steep slopes and a flat surface whose height can be several tens of meters. They were formed using wind and water, thanks to the rivers of ice pass, which at that time divided the flat plain and gave them this shape.

Axial part of the Turgai plateau, called the Turgai gate, is cut through by a wide lodge, there are many lakes with sandy massifs and river valleys.

Beautiful mountain landscapes are located in the south of the Kazakh country fold. In the south of East Kazakhstan region are located the city of Genghistau and have characteristic features and similarities with Central Kazakhstan and Central Asia, there is also steppe vegetation (steppe grass, as well as tavolga and karagana bushes). Kyzylaray is a mountain-bearing granite massif, an oasis in anhydrous semi-deserts.

Orthau and Aktau have specific forms that were formed by the weathering process. These landforms are often found in the southern steppes of Kazakhstan and have forests with small leaves. The south of Turgay is characterized by landscapes of lakes and river valleys, such as Akkol, Zhana-Akkol, Aschytastysor. The Uli-Zhilanshik river has places for nesting waterfowl, as it is fresh and salty.

In the west of country, the Caspian lowland has contact with the Mangyshlak plateau, which consists of low-mountain ridges of Karatau, North and South Aktau and dense depressions. Karagiya and Kaundy have a depth below sea level. Mangyshlak and its plains have a relationship with the flat plains of the Ustyurt plateau, which is limited to cliffs from the Trans-Ural plateau—in the north, the Caspian lowland and the Mangyshlak plateau—in the west.

Betpakdala is an elevated plain in a number of deserts in Kazakhstan.

The eastern part of relief has low mountain, and the western part has flat. The Spurs of the Shu-Ile mountains have spurs of wedging and a high handful as the Kazakh shallow spur.

Relief of the Kazakhstan deserts is diverse and has various forms and has been formed within a long time and depends on the geology of underground complexes.

The Turan lowland, the western part of Betpakdala, the Ustyurt plateau are mainly located on the Turan plate in desert zone. Exits on the surface are in the form of chalk sand-clay strata.

Marly-clay sediments are often found in the north of Ustyurt, in the north of the Aral Sea region and in the west of Betpakdala. These precipitates when coming to the surface protrude in the form of stops and ledges of boards. Ustyurt is composed of Miocene deposits of limestone-marly and clay composition. The Mangyshlak peninsula consists of Triassic and Jurassic sandstones and clays, especially the Karatau and Aktau mountains.

Many geological rocks characterize this desert as a rich land for minerals. In this area, all species and types of deserts are found: sandy, pebble-sandy, loamy, clay tacyrus, loess, crushed stone, rocky, etc. The sharp continental climate of Kazakhstan deserts is characterized by dry air, lack of water resources, etc. (Akiyanova et al. 2019).

The work identified two promising areas for creation of geoparks, they are Central and Western Kazakhstan.

Saryarka is a territory that shows the periods of Earth development and the features of Eurasia history, we can say the natural «geomorphological museum» of relief (Saryarka turistski putevoditel', Saryarka tourist guide 2018).

In 2008, the UNESCO World Heritage List included unique forms of the relief of Nauryzym and Korgalzhyn reserves such as «Saryarka—Steppes and lakes of North Kazakhstan» (https://ru.wikipedia.org/wiki/).

Fig. 4.2 Saryarka (authors)

Saryarka has a length from west to east—1200 km, width in the eastern part—400 km, in the western—up to 900 km. The area, which was formed in ancient times and today is very modified by the natural forces and consists of low mountains of shallow ones and small hills (Fig. 4.2).

Between the small hills of Saryarka there are lakes, basins, depressions and other geological forms and structures.

The western part of Saryarka has a higher surface than the eastern part.

Saryarka consists of magma and sedimentary rocks, for example quartzite, granite, porphyrit of the Paleozoic era.

When forming the terrain in this area, location of rocks and wind process played a large role, in this regard, the mountains have an almost ideal flat surface.

Some sections of the mountains have gentle hilly terrain, where a typical landscape of the semi-desert zone is found with characteristic low hills 50–150 m high, which are surrounded by equal plains.

In the east of shallow settlement are the highest mountains. Such as Kyzylaray—a mountain-hillside granite massif (with a maximum peak of Aksoran 1565 m high), Karkaraly mountains (1403 m), Kent (1460 m), Genghiztau (1300 m), Bayanaul (950 m) (Beysenova 1979).

Slopes of Kyzylaray are covered in the northern part with pine forests, and the southern slopes are less rich in vegetation and rocks.

One of the large basins of Kazakhstan is the Tengiz-Kurgalzhyn depression (304 m), which divides Saryarka into two parts.

Ulytau is located in the northwest of Kokshetau mountain. The Kokshetau mountain does not have high peaks, canopying about 900 m high. The mountains consist of limestone, quartzite and Paleozoic porphyrite.

Slopes of the Kokshetau, Ulytau mountains are greatly changed by the waters and in the intermountain valleys formed beautiful unique lakes that are bordered by pine forests.

Ulytau (1133 m) from the Kazakh language is translated as Great mountain, which consists of many large anticlinals located meridially and consisting of granite.

The slopes of Ulytau mainly consist of shale, sandstone, conglomerate and are strongly dissected by the natural forces. The Region attracts with its birch pews located on a plain with hills composed of clay rocks of the lower Cenozoic (Saryarka turistski putevoditel', Saryarka tourist guide 2018).

Saryarka is rich in various fossils. For example: Karagandy and Ekibastuz coal basins, which provide the regions with coal and are a powerful fuel pool. Rich ores of copper, iron and manganese are well-known for the deposits Zhezkazgan, Karsakpay, Atasu, Sayaksy, and Kounyrat. Rare metals are found at the Zhayrem deposit.

Then there is the process of manufacturing them at the huge metallurgical plants of Karagandy, Zhezkazgan, Balkhash and Temirtau.

Saryarka is located in the central part of Kazakhstan, in this regard, the climate here is sharply continental and in most cases there is an anticyclone and air masses that lead to a dry climate.

Extreme climatic conditions are a limiting factor for the development of some sectors in economic activity, including tourism.

Here, there are many rivers, such as Yesil, Nura, Sarysu, Selety, Shiderti, Tokyrau.

Rivers have spring nutrition and in the spring come out of their banks and can form a wide floodplain.

In summer, on the contrary, the rivers begin to dry out, which lead to the formation of mould and old things, sometimes the river bed dries up, only the river Yesil has a constant flow.

On the territory of Saryarka, unique relief forms are everywhere found that can remind us of fabulous heroes, for example: in Kokshetau—«Camel», «Sleeping knight», «Sphinx», «Zhumbaktas», «Golden eagle», «Okzhetpes», in Karkaraly—«Stone Toad», in Bayanaul mountains Baba-yaga, etc.

These unique rock forms are surrounded by beautiful and unique nature Akiyanova and Bekkuliyeva (2016) and Yegemberdiyeva et al. (2017). Mountains Kokshetau, Karkaraly, Bayanaul, attract many tourists with their picturesque beauties.

Unique landforms have the surroundings of Aktau (Western Kazakhstan).

Aktau is a unique city that has a specific climate and feature in location attracts many tourists and in 1963 was founded as a place with uranium deposits.

The closely located territories to Aktau are known for their beautiful and specific landscapes of the Mangyshlak plateau deserts. Specific climate and wind-water erosion transformed solid rocks and created many miniature mountain chains, small canyons and deep depressions that contain salt water.

Mangyshlak peninsula is located in the east of the Caspian Sea, which is located in the west of Kazakhstan. Since the 1990s, the peninsula has been renamed Mangistau. The Peninsula is rich in oil, unique the natural and historical monuments.

Tuzbair plateau is a rocky area and a salt marsh in the Mangistau region in southeastern Kazakhstan, near the large Ustyurt plateau.

Mangystau area is an outstanding landscape beauty: snow-white salt marshes and limestone mountains against the background of a blue sky and yellowish-brown soil (https://www.tourister.ru/world/asia/Kazakhstan/city/shetpe/placeofinterest/32490).

Tuzbair (Fig. 4.3) is the greatest the natural monument created by the natural forces. In this place, the plateau is steeply torn by a cascade of clay-limestone steps. Such a natural formation is called a chink. The lower stages pass into a vast salt marsh, called soar in local edges.

Terrain is diverse and impresses even experienced travellers: rock cliffs, weathered remains, salt marshes, and steppe. It is worth staying here for at least a day to observe the landscapes under different lighting—in soft morning and in bright daylight, golden sunset light, as well as see the outlines of rocks under the starry sky.

The best time for Tuzbair's visit is the second half of April and May.

Steppe blooms at this time—there are fields of wild tulips everywhere, and the air temperature is kept at a tolerable mark (but the nights are cool).

In summer, the heat here reaches +40 … +45°. In winter, cold winds blow with sands and dust from limestones. In autumn, the weather is also quite mild, but the steppe grasses are dry, the area does not look as picturesque as in spring.

The Main attraction of Tuzbair is the salt marsh of the same name. In Kazakh, it is called «sor», which means a type of salt marsh that does not have run-off and is shallow. Sorrows are formed in depressions with a depth of no more than one and a half meters. During the period of spring melting, the salt marsh is covered with water ironing, in which the sky is picturesque. In summer, the crust is frozen, you can walk unhindered—occasionally the legs will fall into salt deposits to a depth of 10–15 cm. Therefore, to walk on the salt marsh surface, you need to put on rubber boots or other non-washable shoes with high calf. Or alternatively, walk barefoot. Some tourists believe that warm salt crust has a beneficial effect on the body. You can't drive a car on a salt marsh—the weight of even a crust car can't stand it. Every

Fig. 4.3 Tuzbair plateau (own authors)

year, cars get stuck in the marshy area of the salt, which are then very difficult to pull out. You can only drive here with local guides who know proven trails.

The Airakty castle valley is a complex of small remnant mountains that the local population calls Airakty-Shomanai (Fig. 4.4).

Mount Sherkala is a mountain that lonely stands 170 km from the city of Aktau, near the Shetpe village. If you look at Shergaly from a bird's eye, we will see a giant flying saucer, which somehow suffered a torsion (Fig. 4.5).

The Boszhira tract is located on the western outskirts of the Ustyurt plateau—this is a remnant mountain range. It is surrounded by a huge canyon, which repeatedly enhances the impression of contemplation (Fig. 4.6).

Kazakhstan is a modern relict geomorphological park, an open-air the natural museum, which includes: the natural scenic places, for example: Ustyurt plateau, Cretaceous mountains in Mangistau, Aktolagay, chinks and steep ledges of the Aral Sea region, the Karatau ridge, Charyn canyon, the Altyn-Emel mountain range, folded-blue mountains and Altai tundra, as well as the well-known Akzhar collapse.

4.2 Mineral and Thermal Waters and Healing Mud

There is a particular interest for recreation and tourism at the exits of medicinal mineral and thermal waters, mud springs, the natural lakes, etc. Because of them, health resorts, hospitals, sanatoriums and resorts are created.

Fig. 4.4 The Airakty castle valley (https://Kazakhstan.travel/photobank)

Fig. 4.5 Mount Sherkala (own authors)

Fig. 4.6 Boszhira tract (authors)

Currently, Kazakhstan has over 500 mineral water sources and more than 80 lakes with medicinal mud, over 50 health-improving areas (Tourism of Kazakhstan 2015).

In Kazakhstan, approximately 10% of available resources are used for health purposes. Radon and siliceous mineral waters are most widely used in health practices. In the republic there are the following types of resort and sanatorium institutions (Tourism of Kazakhstan 2015):

1. Mud resorts (Zhanakorgan, Moiyldy, Shchuchinsk and others) for the treatment of movement organs diseases, the nervous system, gynaecological diseases, skin diseases.
2. Balneological resorts (Almaarasan, Kapalarasan, Saryagash, Merke, Rakhman's keys, Tauturgen, Akkaiyn and others) for the treatment of diseases at central and peripheral nervous system, diseases of the movement organs, skin.
3. Balneological resorts with drinking mineral waters (Saryagash, Temirlanov, Barlykarasan and others) for treatment: diseases of digestive organs, metabolism, and urinary tract.
4. Climatic resorts (Borovoe, Karkaraly, Aman-Karagay, Mankent, Khan grove, Uralsk and others) for treatment of diseases of the respiratory and upper respiratory tract, diseases of the cardiovascular system, blood system, diseases of digestive organs on functional nature, metabolic diseases, kidney diseases.
5. Resorts for the treatment of tuberculosis (steppe with culinary treatment: Borovoe, Berezovka, Sartugay, Chimgan and others).

Moiyldy is a natural lake with unique healing properties, located 15 km southeast of the Pavlodar city. Artemis saline crustaceans live in the water of the lake, which, as a result of dying, form a valuable silt—the well-known Moiyldy mud. Its composition is unique, the entire table of Mendeleev is collected in it. There are no analogues of this therapeutic mud in Kazakhstan. In addition, the lake is covered with a layer of rapa—water, which is a saturated solution. Rapa is the second important healing component of the lake. Source of underground mineral water near the lake is another component of recovery. This water helps to solve many health problems.

Mineral water «Moiyldy» belongs to the balneological group of waters, therapeutic effect of which, when ingested, is ensured by sodium chloride and sodium sulfate. Sodium chloride promotes the activation of metabolic processes in the body, increasing acidity, enhancing the secretory function of the stomach and stimulating the pancreas. The sulfate ion enters the compound with sodium cations, forming, which are not absorbed in the intestine. Irritating the mucosa of the gastrointestinal tract increases motor function, which causes relaxation. The sulfate ion contributes to the release of salts on heavy metals (lead, mercury) and during occupational poisoning reduces degenerative processes in the liver. Effectiveness of mineral water when ingested is influenced by the time of water in mouth: this mineral water should be taken in small throats for better development by the stomach. Mineral water is recommended for gastric diseases, heart and vascular diseases, gynaecological diseases, diseases of musculoskeletal system.

Thus, the uniqueness of lake Moiyldy, the waters of which help strengthen the human body, contributed to the creation of balneograceatel of the same name here, which has its formation history.

In 1912, one Pavlodar merchant, a certain Mr. Slavtsov, arranged a paid mud clinic on the shore of lake, and in 1922, when Soviet power was established, it was transferred to the Pavlodar «Gorzdrav». Gradually, the resort began to transform, a mud clinic was rebuilt with a throughput of up to fifty people a day.

In 1928, after the sanatorium transferred to the Republican resort administration, Moiyldy turned into a health resort of Republican significance, and its throughput began to increase rapidly and amounted to more than 250 people per day. Buildings were built for the living of patients and the placement of sanatorium services, and a beautiful park appeared.

The highest heyday of the mud clinic «Moiyldy» occurred in the 80s of the last century. In those years, the sanatorium had six sleeping buildings, a summer dining room with an open terrace, sports and playgrounds, a dance floor, a powerful warehouse and much more. On the sanatorium territory, an orchard was laid out, in which there was a town of prefabricated Finnish houses for summer accommodation for holidayers equipped with water supply and sewage. Houses could accommodate both holidayers with families and collectives. There was also its transport workshop with a large fleet of equipment. During its heyday, the sanatorium could receive 1800 people at the same time. Therefore, it was called the largest sanatorium in Kazakhstan and was considered one of the largest sanatoriums in the Soviet Union (http://moiyldy.kz/info/). Now Moiyldy is one of the leading health resorts in Kazakhstan, where people from near and far abroad come with pleasure.

In the Kyzylorda region of South Kazakhstan there is a well-known lake Teresken, containing highly active silt mud and rapa, having a healing effect with their therapeutic external use.

On the basis of modern data and stock materials of the balneology department, mud of lake Teresken are evaluated as therapeutic silt sol saturated low-alkaline strongly and medium-sulfide satisfactory sanitary-chemical state. In its the natural form, it is suitable for applications without preliminary special preparation on vibrating electrodes or other cleaning that reduces the degree of clogging. Prerequisite for use is a satisfactory sanitary-microbiological state and stability of physicochemical parameters.

On the shores of lake Teresken in 1918, the Zhanakorgan sanatorium was built, which is one of the oldest unique mud resorts in Kazakhstan. Geographically, the sanatorium is located between the Syr-Darya river and Karatau mountains in the Kyzylorda region, 1.6 km southeast of the urban-type settlement Zhanakorgan, 180 km from the city of Kyzylorda.

Since the end of the XXth century, drinking therapeutic mineral water has been mined; this is used to treat internal diseases, and has received a decent rating among holidayers and patients. As of 2017, water is estimated as complying with the State standard of the Republic of Kazakhstan—ST RK 452-2002 in the status of mineral the natural medical and canteen drinking water.

Out of the trace elements in this water, the presence of the natural iodine at an active concentration of −100 mg/dm^3, bromine ions at a concentration of 2.85–1.0 mg/dm^3, manganese ions—0.24 mg/dm^3, fluorine ions—0.45–0.60 mg/dm^3, increasing the balneological significance of this water. The presence of organic substances in an active concentration exceeding −5 mg/dm^3 is noteworthy. The studied water can serve as an effective non-drug treatment and prevention of a number of diseases, including digestive organs, musculoskeletal system and is recommended for use in this sanatorium (https://szhanakorgan.kz/page/about/8).

Unique natural components—the healing mud of lake Karabotan and the mineral brine of well No. 12 serve as the basis for functioning of the Atyrau sanatorium. Resort «Atyrau» is considered the oldest health resort in Kazakhstan, founded in 1936. Sanatorium is located in the quiet environmentally friendly centre of city of the same name on the banks of the Ural river in West Kazakhstan.

In 1949, the honoured doctor of the Kazakh SSR A. Z. Sundetov first conducted scientific observations in hospital conditions and revealed a good effect from mud treatment of lake Karabotan. According to the data results received, the health department of the Guryev regional executive committee petitioned for the feasibility of opening a mud sanatorium.

The well No. 12 is located on the territory of sanatorium complex, has been operated since June 1988. Sol saturated sodium chloride brines enriched with bromine and iodine are produced from a depth of 320 m. Such water generally contributes to body health, improving blood circulation and sleep (https://sanatory-atyrau.kz/mineralnaya-voda). Currently, Atyrau functions as a balneograde sanatorium.

In South Kazakhstan, 130 km from the city of Shymkent on the left bank of the Keles river, the Saryagash resort is located. The main property of the resort is mineral waters belonging to the group of nitrogen ectotherms and are an effective non-medicamentous agent for prevention and treatment of a number of diseases outside the exacerbation phase.

The source of Saryagash mineral water was discovered in 1946. In 1946, on the left bank of the Keles river, among the sandy hills, geologists of the SREDAZNEFTERAZVEDKA trust opened hot clear water at a depth of 1400 m, which turned out to be mineral. The same water was brought to the surface and through another well.

In 1949, employees of the Uzbek scientific research institute of spa and physiotherapy named after Nikolay Semashko began research on this mineral water (https://sariagash.kz/istoriya/).

In 1953, the South Kazakhstan health department opened a balneological hospital here. This made it possible to start studying the healing properties of mineral water.

Chemical composition of the natural mineral water is weakly mineralized (mineralization—1.65 g/dm^3) of a weakly alkaline hydro-carbonate sodium composition of low total stiffness (0.50 mmol/dm^3), similar in basic composition to the mineral waters of the Shalkar hydro chemical type (type Ivb ST RK 452-2002). A specific feature is the presence of fluorine ions—3.2 mg/dm^3, a preventive dose of the natural iodine—90–100 mg/dm^3. Content of heavy metals does not exceed permissible standards (mg/dm^3)—copper-0.004; lead-0.019; zinc-0.008 (https://sariagash.kz/lechenie/).

People have become aware of the healing properties of radon waters a long time ago. Because of radon sources in East Kazakhstan, there is a sanatorium «Rakhman keys», equipped with the latest medical equipment. For each patient, the doctor develops an individual treatment program. Specialized medical services of the sanatorium are: hydro- and physiotherapy, mud therapy, thermo therapy, massage, pantology. This allows us to cover a rather wide range of medical indications for which prevention and rehabilitation are carried out here.

Since 1964, the «Rakhman keys» sanatorium has been operating. The very location of the health resort among picturesque Altai mountains within the Katon-Karagay national the natural park attracts many holidayers here. The sanatorium is located at an altitude of 1760 m above sea level, beneath the Holy Belukha (https://rahmany.kz/about).

Water of the Rakhman keys is unusual in its healing power and composition, since the radon content causes optimal effect for the human body.

Outputs of radon sources are also available in South Kazakhstan. Here there is a sanatorium «Merke». The Merke mineral radon water deposit is located 4 km south of the Merke sanatorium, on the northern slope of the Zailiysky Alatau in the picturesque valley of the Merke river. According to the main chemical and balneological indicators, the water of this deposit belongs to the group of medium and high radon fluorine therapeutic mineral waters; in terms of the enrichment level, radon is one of the in the Republic of Kazakhstan and Central Asia. The main component in the physiological influence on body in this water is its high radon index Rn 87–136 nCi/dm^3, nitrogen-gas composition (93% nitrogen), alkaline reaction PH 7.9–9.5, increased fluorine content 9.4–15.0 mg/dm^3, silicic acid 32–35 mg/dm^3, as well as a wide range of therapeutically valuable trace elements: It should be noted that selenium at a concentration of 10–20 mg/dm^3 contributes to an increase in immunity, reduces the risk of developing cardiovascular, endocrine diseases, and removes toxic substances from the human body (https://merke-radon.kz/).

The Okzhetpes health centre is located in the heart of Kazakhstan pearl—on the shores of lake Burabay, the most beautiful of the lakes in Shchuchinsk-Borovoe resort area. There is a unique the natural attraction of the village, the visiting cards of Borovoe—the fabulous rocks of Okzhetpes and Zhumbaktas. In a three-minute walk along a dense pine water side is the legendary glade of Abylay Khan.

Okzhetpes is a climate balneological complex of year-round action.

Currently, the Okzhetpes health centre operates year-round and has a number pool for 173 holidayers and is a health resort for year-round operation.

In 2006, the General Assembly of the World Diving and Climatology Federation in Andorra awarded the Okzhetpes health centre the award «The Best resort in the world» for achievements made in organizing sanatorium and resort treatment and active scientific activities among the World Health Organization (https://okzhetpes-burabay.kz/ru/o-nas/).

Not far from the southern capital of Almaty is the Ak-Kaiyn sanatorium complex, which has a medical and healthy orientation. It is located in the mesmerizing view of the valley at mountains of Zailiyski Alatau, the height of which rises to 1300 m above sea level. Local air is saturated with negative ions, which, like invisible medicines, have a therapeutic effect on the body. They normalize blood pressure, improve blood circulation, stimulate good appetite, normalize metabolism, lower cholesterol and sugar in bloodstream. In general, they have more than seventeen positive influences. The sanatorium got its name due to the nearby birch grove (http://www.ak-kain.kz/sanatorij/).

Healing mineral springs overlooking the Earth's surface, the local population has skilfully used since ancient times. The Kazakh people had their ways of treating mineral waters, they were also described by the well-known Kazakh scientist, historian, traveller Shokan Ualikhanov. Such treatments include:

- the use of silt mud and salted horn in the form of applications, called «Tuz»;
- use of drinking mineral springs, as well as for bathing—«Maden-su»;
- use of hot mineral springs for health and healing purposes—«Arasan».

In Almaty region near the city of Zharkent there are such springs with hot mineral water of radon origin. Since 1967, the Zharkent-Arasan sanatorium has been operating because of these sources.

The reason why, throughout the world, the healing power of underground sources is considered, if not miraculous, then, at least, unique and fabulous, is that it is impossible to synthesize the chemical composition of living the natural mineral water—man still does not have such powerful energy and technological resources to reproduce the most complex and powerful geochemical processes that occur in the Earth bowels.

4.3 Areas of Relict Vegetation

In Kazakhstan, ecotourism is most actively developing in specially protected the natural areas with unique the natural complexes. However, for the development of tourism, including geotourism, territories with areas of relict vegetation are of great importance.

In 11 state national the natural parks, reserves and territories, with active economic activity, places of relict plants growth are concentrated (Table 4.1).

Charyn ash grove. **The total** area of forest cover is 2700 ha, the length is 26 km, the width is up to 2 km (Ivashenko 2009). The ash grove is located in the Charyn state national the natural park, Almaty region.

Sogdian ash (Fraxinus sogdiana Bunge), the olive family, (Oleaceae) is a tree with an average height of about ten meters, the thickness of trunk can reach 10 «girth». Individual specimens reach a height of 30–35 m with a boot thickness of

Table 4.1 Areas of relict plants in Kazakhstan

No.	The natural object	State status	Year of establishment	Area, ha	Location
1	Charyn ash grove	Charyn state national the natural park	2004	127,050	Almaty region
2	Sinegorsk fir grove	State the natural monument	1981	137	East Kazakhstan region
3	Relict array	Kokshetau state national the natural park	1996	182,000	North Kazakhstan region
4	Relict «Silver wood»	The natural monument of Republican significance	–	250	North Kazakhstan region
5	Relict wood «Zhanazhol»	The natural monument of Republican significance	–	16	North Kazakhstan region, Zhambyl district
6	Pinery	The natural monument of Republican significance	1905	26.0	North Kazakhstan region
7	Ribbon pine wood	State forest the natural reserve	–	662,000	East Kazakhstan region
8	Chinturgen relict moss spruce	The natural monument of Republican significance	1968	887	Almaty region Kairak river gorge (left tributary of Turgen river)

more than 2 m, and their crowns spread 18–25 m in diameter. Ash grows slowly under the natural protection of canyon walls, lived for several hundred years, there are separate trees that are already more than 700 years old.

The ash grove is the only island of relict forest among the CIS countries preserved from the Neogene period (25 million years). Sogdian ash has long become a relic variety and a rarity. In addition to ash in the canyon of Sary-Togai, you can find turanga (diverse poplar). Trees with a curved trunk (and sometimes multi-trunk) more than 20 m high, with leaves of different shape and size with a bluish-green colour. One of the features of this species of plant is the released salt juice in the affected places, a kind of «tears». Around the world, Turanga is actively used for landscaping desert areas, thereby ennobling settlements.

A separate charm and attractiveness to the landscape of ash grove is added by the typical rocky desert of the canyon, mammals of inhabitants (fox, mallard, steppe chorem, reed cat, wild boar, ondatra, roe deer, hare) and birds (black stork, tuvik, white-tailed orlan, pheasant, brown pigeon, eagle owl, scops owl etc.)

Maximum tourist load is noted on weekends and in the summer, when the number of visitors reaches 200–250 people. On average, daily attendance is small. Recommended visit periods: April–June; August–October.

Sinegorsk fir grove is located in the East Kazakhstan region, on the northern slope of Mount Medvedka, borders from the north and east with lands of the Ulansk administrative region, from the south and west—with the Sinegorsk forestry and covers an area of 137 ha (Nikiforov 2005; Chernetskiy 2008). Sinegorsk fir grove has the status of a state the natural monument, created to preserve the redwood of relict boreal fir with the embossing of pine and aspen, which has great scientific and cultural-cognitive significance.

The nature monument includes about 400 species of vascular plants, out of which 38% of the area is woody vegetation, represented by Siberian fir, aspen and wart birch.

Fir trees grow among rocks in a moistened log, forming in places both clean and mixed plantations. There are trees over 15 m with a trunk diameter of 20–30 cm.

On the territory of the natural monument there are red book plants: Siberian juniper, Altai rhubarb, peony, swimsuit and others. In the animal world of the fir grove there are wild boar, elk, maral, wolf, hare, fox, squirrel, bear and many other animals. Ornithofauna is numerous, among them there are red-book representatives: eagle-golden eagle, falcon-baloban, eagle owl, black stork. The maximum tourist load is noted in summer, when the number of visitors reaches 100–150 people. On average, daily attendance is small.

The relict massif is located on the territory of the Kokshetau nature park, in the Zerendy district of Akmola region and the Ayyrtau district of North Kazakhstan region. The relict array has an area of 2.0 ha (https://gnppkokshetau.kz). Here is a rare herbaceous species of perennial eagle fern. Fern grows under the canopy of coniferous forest, dominating the grass cover. Massif is a favourite place to visit, both tourists and the local population, tourist trails and routes pass to it. On average, daily attendance at the facility is small, reaching up to 200 people on weekends.

Relict «Silver wood» is located near the village of Bolshaya Malyshka, North Kazakhstan region, with an area of 243 ha (Ivashenko 2009). The relict wood is a natural monument of republican significance. The rarest species of trees, whose age is 100 years, surrounded by birches are attracted by their uniqueness and beauty. On the territory of the boron grows lingonberries, strawberries, and bony. Moss and lichens are adjacent to valeriana and windmill. Attendance at the facility is on average small, in the summer period it is 250–300 people.

Relict wood «Zhanazhol» is located on the territory of North Kazakhstan, 3–4 km from the village of Zhanazhol. The total area of relict wood is 9 ha (Ivashenko 2009). Relict wood is a unique the natural monument of great scientific value.

Tourists are attracted by the natural plantings of relict pine wood. Common pine dominates, the age of which reaches more than 200 years. The average height of trees is about 20 m, the average diameter is about 30 cm. Here you can find various types of mosses and lichens, oryak fern, medicinal, birch and aspen bough.

Wood Zhanazhol is in the physical and geographical zone atypical for it, so it needs a particularly careful attitude. A favourable factor is the distance from large settlements.

Pine wood—the natural monument of republican significance, is located on the territory of North Kazakhstan. The total landing area is 26.0 ha (Ivashenko 2009). Pine wood is an artificial plantation on ordinary chernozem. Currently, trees are characterized by high-quality trunks. The average diameter is 18–24 cm, 16–20 m high.

Ribbon pine wood is located on the territory of East Kazakhstan, stretches in stripes along the sandy ancient alluvial deposits of the Irtysh river. It has the status of a state forest the natural reserve.

The wood has great sanitary and cultural and recreational significance. Clean air, saturated with resins attracts many tourists and people who want to improve their health, here you can relax and gain strength. On the territory of pine wood there is a regional tuberculosis dispensary (Chernetskiy 2008).

Chinturgen relict moss spruce is a natural monument of republican significance, located on the territory of the Ile-Alatau national park, in the gorge of the Kairak river, Almaty region. Total area 887 ha (Ivashenko 2009).

Currently, the relict moss spruce is the only place on Tien Shan where such spruce trees have been preserved intact.

The landscape of a moss spruce is similar to the forest tundra zone—the remains of a glaciation massif, low air and soil temperature, moss cover, up to 50 cm high. Relict spruce trees grow at an altitude of 1850 m.

Tourists are given the opportunity to walk along a complex ecological trail: the narrow Chinturgen gorge, many showers and felled spruce.

4.4 Future Expectations of Geoparks Development in Kazakhstan

Since the early 1990s, in developed countries, the emergence of a movement to create geoparks began, which over time covered the entire planet. Since 2002, there has been a special UNESCO programme to support the Global Network of National Geoparks. Since 2004, the organization has been developing the so-called geotourism (https://www.kazinform.kz/ru/geopark-sozdadut-v-mangistauskoy-oblasti_a3055556). By the beginning of 2020, the World Network of National Geoparks included 147 geoparks from 38 countries.

Geopark—a region with a specially protected status, on territory of which the geological Earth's history is clearly revealed, formation of local landscapes and deposits of minerals, mineral and thermal waters, sources of healing mud, fossil remains of prehistoric animals have been preserved. Geoparks are special zones of non-living nature, formation history of which has scientific and cognitive significance relative to the Earth's history. On specially laid routes, equipped with signs and travel maps, tourists and excursionists can travel.

Kazakhstan, like Central Asia as a whole, is rich in all kinds of unique geological and geomorphological objects. However, despite such great potential, there is not yet a single geopark in Central Asia. It is necessary to move away from the classical understanding of geology purpose not only for the needs of the mining industry, but also to promote aesthetic values and knowledge of the Earth's history through geological objects.

UNESCO's global geoparks support sustainable development initiatives through local initiatives, especially the development of ecological tourism. In Kazakhstan, with the correct use of geoparks as objects of geotourism and ecotourism, it is possible to make a significant contribution to the development of the country's economy. In addition, geoparks help monitor and raise awareness of the natural resource management, climate change and risk of the natural disasters.

Work on the study of geological heritage has been carried out in Kazakhstan for a long time. A significant contribution to the geology study of the Earth's structure on country territory was made by scientists I. Fishman, Y. Kazakova, who did a great job of creating a database of 500 objects, could release photo albums «Geotopes of Kazakhstan» and «Millions of years before the Silk Road».

In Kazakhstan, there are prerequisites for the creation of geoparks not only on territory of the Mangistau and Shchuchinsk-Borovoe resort zone, but also in Almaty and East Kazakhstan, Karagandy and Pavlodar regions, that is, in all regions where there are unusual landscapes.

For example, in Pavlodar region there are unique forms of the Bayanaul relief, sources of mineral waters and healing mud of Moiyldy, a palaeontological the natural monument of Kazakhstan «Goose flight».

The «Goose flight» is a palaeontological burial ground within the city of Pavlodar, in its northwest (the right steep bank of the Irtysh river), on an area of 2 ha. This is a high sandy cliff of the Irtysh river, where, at a depth of 6–8 m from above, bones of fossil animals of Neogene age were preserved—antilops, giraffes, deer, sabre-toothed tigers, ostriches, turtles and other belly, which lived here 12–14 million years ago. The thickness of the layer containing fossil bones is three meters (Chigarkin 1980). Burial of the «Goose flight» is classic of animal bones, it is world well-known. In 1971, the burial site was declared the natural monument of republican significance.

Excavations of the richest burials of the ancient animals remains concerning so-called hipparion fauna were carried out by Soviet scientists from Moscow—Yu. A. Orlov and E. D. Shlygin in the late 1920s. Several thousand bones remains of ancient (ancestral form of modern African fauna) mammals of Pliocene-Miocene age were discovered. In particular, the bones of a three-toed hipparion horse that lived 12–14 million years ago, as well as other ancient animals—a rhino-hiloteria, hyena, sabre-toothed tiger-mahyrod, giraffe, turtle, birds were found. Today, there is no such the natural monument in the city limits anywhere else in the world. Unfortunately, all the finds found were taken away by scientists, since at that time there was no museum of local lore in Pavlodar, there was no scientific organization that could ensure the study and preservation of palaeontological remains. The begun work of A. Yu. Orlov in 1960 was continued by an expedition of the

Department of Palaeobiology from the Institute of Zoology at the Academy of Sciences of the Kazakh SSR (https://silkadv.com/ru/node/533).

Because the monument is located nearby of cities, it is the most visited place of residents, guests and tourists of Pavlodar. The «Goose flight» the natural monument is the largest burial ground for animals of the Neogene era, a world-well-known reference monument, according to which all the now studied burials of extinct animals of that time are compared.

Thus, the territory of Pavlodar region can be attributed to one of the promising for the creation of a geopark.

The strength of Kazakhstan is also that in many places, in the same Zailiysky Alatau, geoparks can be adjacent to the natural parks.

To date, to develop sustainable tourism according to the geopark project, four large roots have been prepared in the Altyn-Emel national park: two excursions towards the ridges of Zhongar Alatau—Big and Small Kalkanov with a visit to the Singing Barkhan, the Dinosaur Trail, as well as trips to the mountains of Katytau and Aktau. Today, work is underway to create tours to the Sholak mountains and the Ungirtas cave in the Kaskyr sai gorge.

Leadership of the national park, for development of the «Dinosaur Trail», coming from the Chokan Valikhanov spring between the two slopes of the Kalkans, in the near future it is planned to give some kind of entourage with models of prehistoric animals in the Mesozoic era. Their prints back in the 1920s were discovered here by Russian scientists and transferred to the Palaeontological Museum in Moscow.

Another route—«Lunar Landscape»—runs along the mountains of Aktau, which are considered the most beautiful and impressive place of the national park. One of the largest palaeontological sediments of the Cenozoic era is located here. Tourists will be able not only to see the remains of ancient flora and fauna, but also to independently find fragments of mammalian bones and turtle shells.

In the mountains of Katytau, guests expect a meeting with very bizarre volcanic formations. This is another silent witness to the planet history. According to historical data, in ancient times there were volcanoes, not inferior to such as Vesuvius and Etna in Europe, also Avachinsk in Kamchatka.

The Akzhar mountains in Karagandy region and the Kiin-Kerish tract in the East Kazakhstan region are easily amenable to comparison with the landscapes of Mars. Mixture of gypsum and loam with the smallest crystals of amethyst gives the Akzhar mountains reddish-purple shades, the colours of which resemble the landscapes of the «Red» planet.

The tract Kiin-Kerish in the East Kazakhstan region, is also located away from civilization, 120 km from the small town of Kurchum, from the regional centre of Ust-Kamenogorsk—about 300 km. Sediments of mottled, mostly red clays, eroded by water and blown by all winds, form a bright bizarre landscape here. According to scientists, this place was formed in conditions of the hot and humid tropical climate of the Mesozoic, when dinosaurs dominated the Earth. By the way, there is another the natural monument—Shekelmes. This is the Kiin-Kerish microanalysis. There are even caves and whole mazes of clay. Nearby on the sandy shore of lake

Zaysan you can install a tent town (https://Kazakhstan.travel/publications/ru/98/Kazakhstan-landscapes-from-a-different-planet).

If you continue to compare interplanetary colours, you can compare the natural shape of the natural object Bozzhyra and the landscape surrounding it on Ustyurt with the planet Uranus, which looks sky blue in telescope. Bozzhyra, translated from Kazakh, means grey-blue land (depression). On a clear day, the area seems shrouded in grey and blue fog.

The main objects that most often interest tourists with cameras at the Bozhyra facility are two limestone peaks, which are popularly called Fangs for their shape. Their height from the base is more than 200 m. And with a narrow «observation deck», not far from the main trail, there is a stunning view of the Bozzhyra valley of the castles.

On the way from the city of Aktau towards Bozzhyra (far from about 300 km), tourists can meet another interesting «alien» object—Mount Bokty, which remains about 160 m high. The mountain is symmetrical and from different sides takes the form of either a trapezoid or a triangle (https://Kazakhstan.travel/publications/ru/98/Kazakhstan-landscapes-from-a-different-planet).

Another interesting place to visit tourists on Ustyurt in the Mangistau region is the Valley of Balls in the tract Torysh. This place is well-known for thousands of round stone formations randomly located on this territory. Radius of some of them reaches three meters. It is correct to call these balls nodules. From Latin, this means «thickening». That is, simply put, when various elements and rocks accumulate around an object or organic life, thus nodule grows. Such spectacular nodules as on Mangyshlak have not yet been found in other regions of Kazakhstan.

There is also a route «Lunar Landscape» with a length of 2.3 km in the Aktau mountains. Tourists are invited to visit the «plaster canyon» with the study of tectonic disturbances and traces of volcanic activity. By the way, geologists call these formations a «dormant volcano», which in a thousand year may wake up. Another route runs along the ridge of the Aktau mountains, from where amazing views open. Tourists can see not only the remains of ancient flora and fauna, but also independently find fragments of the modern mammal's bones.

In the Aktobe region there is a Cretaceous plateau Aktolagay, which fascinates tourists with its «alien» landscape. A real find for a tourist can be the search for ancient artefacts of animal origin, for example, the teeth of an ancient shark that once lived in these edges. The Cretaceous plateau Aktolagay stretched from south to northeast, on the border of the Atyrau (small) and Aktobe (most of) regions. Parallel to the chalk plate, the Emba river flows from the western side. On the one hand, Aktolagay is bordered by the sands of Togalay, on the other hand, the sands of Alshinsay are supported by the ridge.

Scientists and enthusiasts travel from all over the world to the southern slope of the Aktolagay plateau. Here, from the foot of several tens of meters up, sheer chalk walls rise, turning into snow-white ovals. Multicoloured layers of Paleogene rocks darken above: beige, greenish, grey and brown, ending in a layer of shell. Today, the Aktolagay South plateau is a geological and geomorphological object of the State Reserve Fund of Republican and International importance (https://Kazakhstan.travel/publications/ru/98/Kazakhstan-landscapes-from-a-different-planet).

Seventy kilometres from the city of Balkhash in Karagandy region there is an unusual the natural formation in form of a mountain called Bektau-ata. Mountainous terrain in middle of the Kazakh steppe, with a radius of about 5–7 km, with small lines. Granite, like a layered mountain in middle of the Baltic steppe, has a pleasant pinkish tint. The best time to visit is spring and autumn. In the golden port there is especially beautiful. Combinations of yellow-green foliage, light-brown stone and blue sky create incredible landscapes. For a wild tourist, hard-to-reach peaks, caves, waterfalls, reservoirs, forests (true small but picturesque), rare animals and plants are available. There is only one—service, small tour base («Saryarka» and the former children's camp «Mountain») can offer only modest accommodation according to Soviet standards.

Thus, on the territory of Kazakhstan there are the natural prerequisites for the creation of geoparks and geotourism development.

References

Akiyanova FZ, Bekkuliyeva A (2016) Karst landforms as the objects of tourism development in Mangystau region. In: The 33rd international geographical congress, 21–25 Aug 2016, Beijing, p 1571

Akiyanova FZ, Nurmambetov EI, Yegemberdiyeva KB, Temirbayeva RK, Tokzhanov OA (2019) Mangistau—Zemlya, ocharovyvayushchaya voobrazheniye (Mangystau—land that enchants the imagination). Institute of Geography, Almaty, p 304

Beysenova AS (1979) Issledovaniya prirody Kazakhstana (Research of the Kazakhstan nature). Alma-Ata, p 248

Chernetskiy VY (2008) Osobo okhranyayemyye prirodnyye territorii Vostochnogo Kazakhstana (Specially protected the natural territories of Eastern Kazakhstan). In: Ekologicheskoye obrazovaniye v Kazakhstane (Ecological education in Kazakhstan), № 4, pp 20–22

Chigarkin AV (1980) Pamjatniki prirody Kazahstana (The natural monuments of Kazakhstan). Alma-Ata, p 143

Ivashenko AA (2009) Zapovedniki i natsional'nyye parki Kazakhstana (Reserves and national parks of Kazakhstan). Almaty, p 284

Nikiforov V (2005) Sinegorskaya pikhtovaya roshcha (Sinegorsk fir grove). Rudnyy Altay, № 127–128, p 8

Saryarka turistskiy putevoditel' Saryarka tourist guide (2018) Karagandy, p 74

Woodward D, Ivanova N, Yegemberdieva K, Akiyanova F, Fishman I (2018) Mangistau aspiring geopark (Kazakhstan). In: 9th ProGEO symposium. Geoheritage and conservation: modern approaches and applications towards the 2030 agenda, Checiny, pp 86–87

Yegemberdiyeva KB, Temirbayeva RK, Kelinbayeva RZ, Orazbekova KS, Khen AP, Yushina YA, Gaynulla SA (2017) Ekspeditsiya «Unikal'nyy Mangystau» (Expedition named Unique Mangystau). In: № 4 Questions of geography and geoecology. Almaty, pp 77–85

Yerdavletov SR (2015) Turizm Kazakhstana (Tourism of Kazakhstan). Bastau, Almaty, p 520

https://ru.wikipedia.org/wiki/

http://ru.nomadic.kz/tours/sightseeing/kyzylarai-the-highest-the-oldest

https://24.kz/ru/news/polezno-znat/item/116360-25-udivitelnykh-mest-Kazakhstana

http://moiyldy.kz/info/

https://szhanakorgan.kz/page/about/8
https://sanatory-atyrau.kz/mineralnaya-voda
https://sariagash.kz/istoriya/
https://sariagash.kz/lechenie/
https://rahmany.kz/about
https://merke-radon.kz/
https://okzhetpes-burabay.kz/ru/o-nas/
http://www.ak-kain.kz/sanatorij/
https://gnppkokshetau.kz
https://Kazakhstan.travel/publications/ru/98/Kazakhstan-landscapes-from-a-different-planet
https://tengritravel.kz/my-country/geopark-dinozavrami-poyavitsya-natsparke-almatinskoy-oblasti-387009/
https://express-k.kz/news/ekskursiya/chto_meshaet_sozdaniyu_geoparkov_v_Kazakhstane-154252
http://www.mirp.kz/rus/articles/8/495
https://www.kazinform.kz/ru/geopark-sozdadut-v-mangistauskoy-
https://silkadv.com/ru/node/533

5 Evaluation of Tourist-Resource Potential on Ecological Tourism at the Scale of Kazakhstan Regions

Recently, the organization of recreation for local people in general and concerning tourism, particularly, as a means of restoring the ability to work, health and strength of people, as well as their development, have acquired great importance. The Person should consider tourism today as an active form of leisure time use. It is an excellent form of comprehensive renewal: active, live, interesting, enriching a person's aesthetic rest, which fully restores their physical and mental condition.

For people living in conditions of growing urbanization and urban agglomerations, isolated from nature, it became a massive desire to temporarily change the situation, thus satisfying the need for rest and restoring work capacity. As a result, tourism has become a massive social phenomenon.

Leisure time organizations precede the identification and study of tourist and recreation resources, which are completed with their assessment. Following that, the rules of operation are determined and protection mode is established.

One of the most important tasks of tourism geography is an appropriate functional orientation on recreational use of territory. As it is known, evaluation in recreational geography is a complex process of interconnections between subsystems and elements on tourist and recreational systems (TRS), particularly, between the subsystems of "the natural complex" and "group of tourists". Each recreational activity and each category of holidaymakers imposes their requirements on the natural complexes and service sector. Therefore, when determining the study objectives, it is necessary to indicate the object and subject of evaluation, which constitutes the first stage of evaluation. The object of evaluation, is often the natural complexes of certain rank, and subject is resting people. Identification of the indicators list considered on the assessment, and finally, the development of rating scales for individual indicators, constitute the subsequent stages of assessment (Theoretical foundations of recreational geography 1975).

It is known that the natural environment plays a critical role in tourist and recreational zoning. However, tourist-recreational areas can also be formed because of the natural resources that are catastrophically unfavourable for recreation and tourism development, as, for example, in certain areas of Western Kazakhstan

K. Iskakova et al., *Ecological Tourism in the Republic of Kazakhstan*,
https://doi.org/10.1007/978-3-030-77462-2_5

(south and southeast part). The Las Vegas, a world-class tourist centre, is located in the Nevada desert. Origin and development can be defined by the point that tasks of developing the territory, of course, dominated in determining area function (Musabaev and Muldagaliyeva 2012). Consequently, formation of centres and areas for recreation and tourism development is largely determined by tourist and recreational needs of territories.

The main direction of recreational assessment on the natural complexes, which considers the psychological, physiological and technological aspects, as previously indicated, is the method of evaluation.

In our case, zoning method was used in 5 regions of Kazakhstan—West, North, East, South and Central, on which a comparative assessment was carried out by the following points:

- evaluation of a fixed set of potential components (number of the natural landscapes, the natural resources, the natural objects, etc.);
- calculation of the final grade quality scales in their point form;
- required number of assessment objects was involved in comparative evaluation;
- territorial framework of comparison was clearly indicated, administrative divisions in the regions were taken.

5.1 Tourist and Recreational Potential of West Kazakhstan

The natural recreational resources. To assess the tourist and recreational potential of the region, for the purposes of recreation and the development of tourism, as well as recognition and characterization of native components, girdling of this territory was hold in accordance with the degree of the attractiveness for tourists on native landscapes by evaluation method.

Evaluation of the natural complexes in the region for recreation and tourism organization is carried out based on the landscape map of the Kazakh SSR (1979), according to the administrative districts of West Kazakhstan region. In the process of assessing landscape types, each of landscape-forming factors, based on the presence of necessary recreational properties, was assigned 1 point.

The most important factors that were considered in the recreational assessment of landscapes were:

- relief, its morphometric properties affecting the recreational specialization of the region;
- climate (duration of favourable period, and number of favourable days);
- water as a factor that determines conditions of swimming, spa treatment, fishing tourism, influencing the choice of transportation means;
- vegetation, in particular forests, as a factor of aesthetic perception, having a great cognitive value;
- animal world (hunting—game animals).

Additional points were given for each object that is of interest in tourism and recreation, especially objects of nature reserve fund because unsustainable development concepts of national programs, an important place belongs to specially protected areas for preservation and sustainable usage of biological diversity. Since these territories in the system of developed concepts, concerning sustainable development must perform specific functions and are as following (Yerdavletov 1999):

- monitoring cores of native and native and anthropogenic;
- reference objects of integrated learning, working, development and enhancing geographical shell;
- places of inner and aesthetic upbringing, learning, relaxation and tourism.

Consequently, preservation of various ecosystems, rare species of flora and fauna based on protected the natural areas is possible by organizing ecological, educational and other tourism types, aimed at both preserving biodiversity and obtaining certain benefits.

West Kazakhstan is the largest among economic regions of a republic and, accordingly, has rather diverse the natural conditions. Variety of native conditions in the area recognizes wealth of vegetation and fauna. Many kinds of vegetation and animals due to their rareness, endemicity, and features listed in the Red Book.

In accordance with the map named "Nature reserve fund of the Republic of Kazakhstan", 55 objects are noted within West Kazakhstan (17 existing and 38 promising) (Report on research work 2013). Among them are the Ustyurt state reserve and other objects of the natural reserve fund. Thus, concentration of listed objects in all the natural zones and in Mugalzhar mountains can contribute to uniform tourist development of the territory and conducting large-scale research projects.

As a result of the research, a map scheme named "Zoning of the West Kazakhstan territory in accordance with the degree of tourist attraction on native landscapes" was compiled.

In accordance with the applied methodology, the most attractive native landscapes were identified by administrative districts of West Kazakhstan territory. The main goal of tourist zoning within the administrative-territorial units (districts) is convenience for future planning and management.

According to the methodology of Yerdavletov (Report on research work 2013), the following categories of tourist attractiveness assessment (degree) on the natural landscapes were adopted:

(1) Very favourable—3 points;
(2) Favourable—2 points;
(3) Unfavourable—1 point;
(4) Adverse—0 points.

Analysis of the assessment showed that the landscapes of first category (neighbourhoods of the cities such as Uralsk, Aktau, Zelenov, Akzhayik, and Terekty districts in West Kazakhstan region, Kargaly, Martuk, Mugalzhar districts of Aktobe region, etc.) have peculiar orographic features, significant water objects (the Ural and tributaries, the Caspian Sea), diverse vegetation (floodplain forests, birch-aspen pegs, alders, willow trees, roses) and animality (deer, hares, and so on) are the most attractive spine and pictorial. Most of the nature monuments in regions are concentrated here. In this area, holiday in all main areas for almost all categories of tourists may be organized.

This evaluation unit includes the following types of landscapes:

(1) wide floodplains of large rivers, composed of sand, occupied by forb-grassy meadows with tree-shrub vegetation (the Ural and Kargaly rivers);
(2) elevated hilly-rolling plains composed of volcanogenic, sedimentary rocks and crystalline schists, shrub-quill-fescue and sometimes tree vegetation (Ural-Ilek interfluve);
(3) lowlands are ridge-riddled with shrub—wormwood—shrub vegetation (Mugalzhar), etc.

The second category (neighbourhoods of the cities Aktobe, Atyrau, Borly, Taskaly (West Kazakhstan region), Karakiya, Tyupkaragan (Mangystau region), Kurmangazy (Atyrau regions), Will river, Kobdinsk, Khromtau, Aytekebi (Aktobe region)), form landscapes contribute to organization of different travel length. These landscapes are described by: the existence of exceptional geological and geomorphological objects in the relief (geological parts, etc.); numerous rivers, wells; existence of birch-aspen fids, steppes and other botanical objects, together with plenty game animals (Report on research work 2013; Geography of domestic tourism of Kazakhstan 2014).

This category includes the following types of landscapes:

(1) denudation low-flatten plains, composed of crystalline schists with shrub-feather-fescue vegetation;
(2) floodplain of Will river with forb-grassy meadows etc.

The third category (Syrym, Temir regions (Aktobe region), Shyngyrlau (West Kazakhstan region), Algyn, Makhambet (Atyrau region), Irgiz, Indery, Mangistau, Beyneu (Mangystau region)) consists of landscapes with not beneficial conditions for organizing recreational activities for some species (for hunting, recreation fishing, education and medicinal tourism). These landscapes are characterized by the presence of unique geomorphological and geological areas in the relief (chalk mountains, etc.); a significant number of favourable days for recreational activities; mineral water deposits and manifestations of therapeutic mud.

This category mainly includes such landscapes as:

(1) plains with wormwood-stipa capillata—agropyron fragilis vegetation;
(2) tablelands, mesa plateau with agropyron fragilis and grass-wormwood vegetation;
(3) alluvial flat plains with anabasis salsa and black-wormwood vegetation, etc.

Landscapes with unfavourable conditions for recreation and tourism are concentrated mainly in semi-desert and desert zones and in some other parts of a region. For example, the Baiganinsk district of Aktobe region is unfavourable for winter recreation development. However, the natural complexes of this category are suitable for hunting, walking, medical and recreational rest of local population (Fig. 5.1).

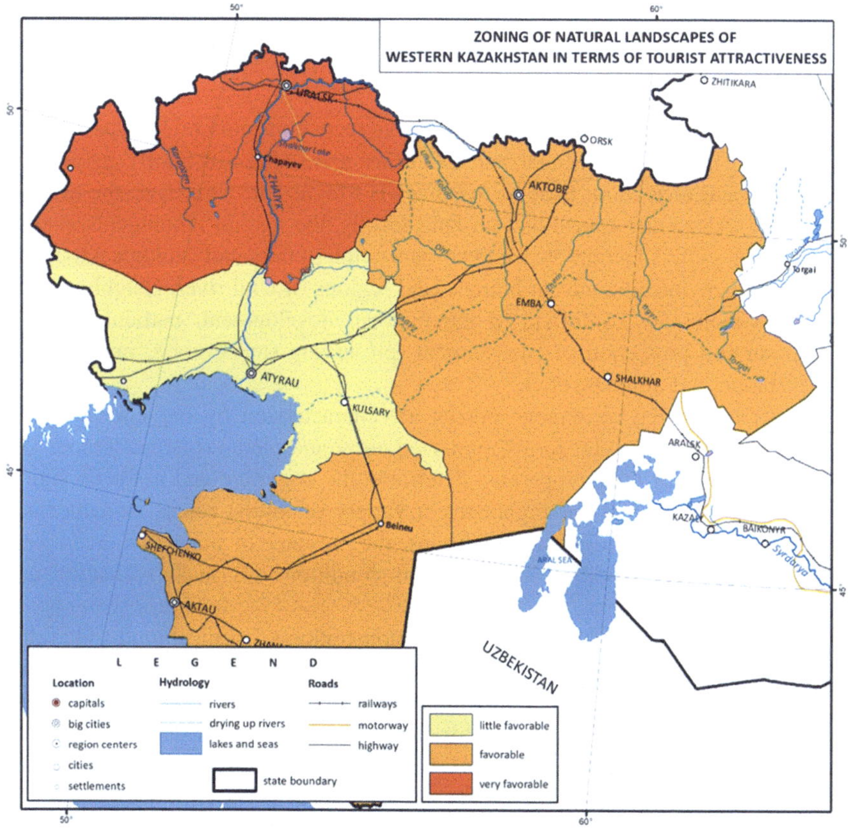

Fig. 5.1 Zoning of the natural landscapes of West Kazakhstan in terms of tourist attractiveness (own authors)

This category includes:

(1) mesa residuals (with turtkul) plateaus with wormwood, feather-grass and spruce vegetation;
(2) reservoir plains with wormwood, feather-grass and spruce vegetation;
(3) aeolian hilly-ridge plains with cereal-wormwood-agropyron fragilis vegetation, etc.

So, landscapes with very advantageous conditions for the development of tourism and recreation with 13 and 12 points; favourable landscapes—10 points; less favourable—by 8 points, respectively; unfavourable—6 and 6.5 points, accordingly.

Analysis of native context for tourism and recreation development in regions and landscape classification which is in accordance with the degree of tourist attraction appears as:

1. West Kazakhstan region has high recreation potential. And it was not deeply studied so, as a result, is nearly not used in tourism and recreation.
2. Mountainous landscapes, vale and to some extent steppe sights have a high degree of tourist attraction.
3. The most advantageous and tempting native recreation sources are typical, as an example, in Kargaly, Mugalzhar and other areas that are poorly evolved in tourism and recreation. Base of resources is Mugalzhar, Ural river, the low-hill flatland, Alexander caves, bottom land forests, the natural monuments, and so on. The region is favourable for the organization of flora and geological tours, as well as recreational zones in bottom land regions of Ural river. Neighbourhood with Russia may contribute to international development, particularly, eco-tourism and protection of environment (preserving steppe areas, rare types of vegetation and animals, etc.).
4. Favourable native recreation sources are characterized by ruggedness of the relief, presence of water forms, protected areas and picturesque landscapes and other objects (or their properties) determine the cognitive and aesthetic significance of these areas for organizing a variety of tourist routes in nature and extent. Weather and climatic conditions, the presence of mineral sources, vegetation and water bodies are important conditions for mass recreation and treatment development.
5. Regions with mineral water deposits, demonstration of medical mud and aridity of climate, together with other factors, is the basis for the development of healing tourism facilities, which are unfavourable native recreation sources. Turgai state reserve, meteorite crater Zhamanshin and a number of other objects favour the development of cognitive, recreational and sports tourism. Organization of special routes is possible in sparsely populated and non-transport territories.
6. Adverse native recreation sources are fairly widespread. Meanwhile, above mentioned regions are of a great interest for science, learning and adventurous tourism and different types of recreation (for example, hunting for animal's behaviour, etc.).

In addition to zoning the territory of West Kazakhstan in accordance with the degree of tourist attraction on native landscapes, authors researched native recreation objects to indicate the functional structure of regional perspective tourism.

According to the results, the summary is that Western Kazakhstan territory is advantageous for developing tourism and recreation, as it has quite diverse and significant native recreation sources.

Socio-economic (anthropogenic) prerequisites for recreation and tourism development. West Kazakhstan is the largest region in the Republic of Kazakhstan, located at the junction of world two parts, bordered by 3 CIS countries, and is connected with Iran and Azerbaijan through the Caspian Sea.

It is known that ethnographic objects, folk crafts and handicrafts, which can contribute to the broadening of tourists, form an integral part of cultural and historical heritage (Report on research work 2013; Geography of domestic tourism of Kazakhstan 2014). For example, it was found that the specific features of folk architecture in North Ustyurt (these are mainly monuments of the Kazakh genus karakta-tabyn) are well expressed in the Big Zhybysky complex, etc.

Architectural monuments on the territory of Western Kazakhstan region is a unique phenomenon, which requires to be increased attention from public and state organizations for their further research, protection, conservation and restoration (Report on research work 2013).

Main forms of tourists acquaintance with the culture of people can be holding music festivals, concerts; sale of disks and video materials with records of national music.

In general, development of culture is relevant and aimed at preserving, strengthening the independence, sovereignty and identity of people.

Peculiarities of national cuisine, range of dishes, their quality often remain in the memories of tourists not only about leisure, but also about country (Report on research work 2013; Geography of domestic tourism of Kazakhstan 2014).

Tourists, especially foreign ones, would be interested in becoming acquainted with the culture and history of not only the indigenous population, but also other people living in the region. For example, in Aktobe region, considering ethnographic features, it is possible:

(1) organization of excursions in Aktobe city with a visit to national cultural centres, churches and other objects (including familiarity with national cuisine in restaurants ("Ochag", "Urartu") and cafés);
(2) organization of tourist route "Aktobe city—Bulgarian village (Algyn district)", since the locality is the second district after regional centre of compact Bulgarians' residence in region. Here there are several Bulgarian families live, who have preserved their original culture.

Thus, the population of West Kazakhstan, representing various nations, has a rich historical and cultural (national) heritage and other socio-economic recreational resources that contribute to the development of cognitive and other forms of recreation and tourism.

Cognitive and educational value, visual appeal were considered during assessment of the monuments, their uniqueness, typicality among the objects of this type.

There are about 3977 monuments of history and culture which have been considered, and 39 monuments of history and culture has republican significance.

Among 3977 history and culture monuments, 2833 are a property of archaeological monuments, 591—history, 255—architectural and 298—monumental art in accordance with state controllers of protecting historical and cultural monuments in Western Kazakhstan.

Depending on the cultural, historical and artistic value, the monuments in question are subject to protection of republican or local significance. The fame and exoticism are the most important properties of a recreational object.

According to research results, the authors conducted zoning of region territory following the degree of saturation with socio-economic (anthropogenic) tourist and recreational resources and made a map "Zoning of the West Kazakhstan territory according to the degree of saturation with anthropogenic recreational resources" (Fig. 5.2) based on archaeological map of Kazakhstan, the register of historical-cultural heritage in region. At the same time, number of the most attractive and studied objects of 61 and more is noted in Karakiya (Mangystau region), Makhambet (Atyrau region), Burly (West Kazakhstan region), Kobdy (Aktobe region) districts; from 41 to 60—Tyupkaragan, Mangystau (Mangystau region), Akzhaiyk (West Kazakhstan region), Inderskom (Atyrau region), Kargaly, Martuk, Mugalzhar (Aktobe region) districts; from 21 to 40—Bokey orda, Zhanibek (West Kazakhstan region), Beyneu (Mangystau region), Zhylyoi (Atyrau region), Khromtau, Shalkar, Temir (Aktobe region) districts; less than 20—other areas, respectively, received a high (3 points), medium (2 points), low (1 point) and really low (0 point) degree of saturation with anthropogenic recreational resources (Report on research work 2013).

Territories of the first group (3-points) are identified by significant and diverse historical and cultural monuments that are perspective for the development of scientific, religious, business and other types of tourism: in Mangystau (complex of underground mosques, etc.), West Kazakhstan region (necropolis Kyrykoba, etc.), Aktobe (Abat-Baitak mausoleum, etc.), Atyrau region (Saraishyk settlement, etc.).

In the territories of the second group (2-points) following monuments are concentrated: memorial complex T. G. Shevchenko (dugout of T. G. Shevchenko; house-museum of 1853 year; well and bust of T. G. Shevchenko, 1888 years); necropolis Masat-ata X–XIX centuries; in Mangistau region and others.

Basis on formation of educational tourism areas of the third group (1-point) can be unique monuments, as an example, the Khan's bases of Bokey horde, located in Western Kazakhstan region. Bokey horde was established in 1801 by Sultan Bokey. In 1801, the son of Sultan Bokey was born, the future Khan Zhangir. Zhangir is a legendary man, called the Kazakh Peter the First. Zhangir initiated sedentary life in these regions. The village, which began with the house of Khan, established in the Khan's bases, was the first centre of politics, cultural and economic life of Bokey horde. There were complexes called Daumshar, Karasakal, Asan-Kozha, as well as a number of other objects.

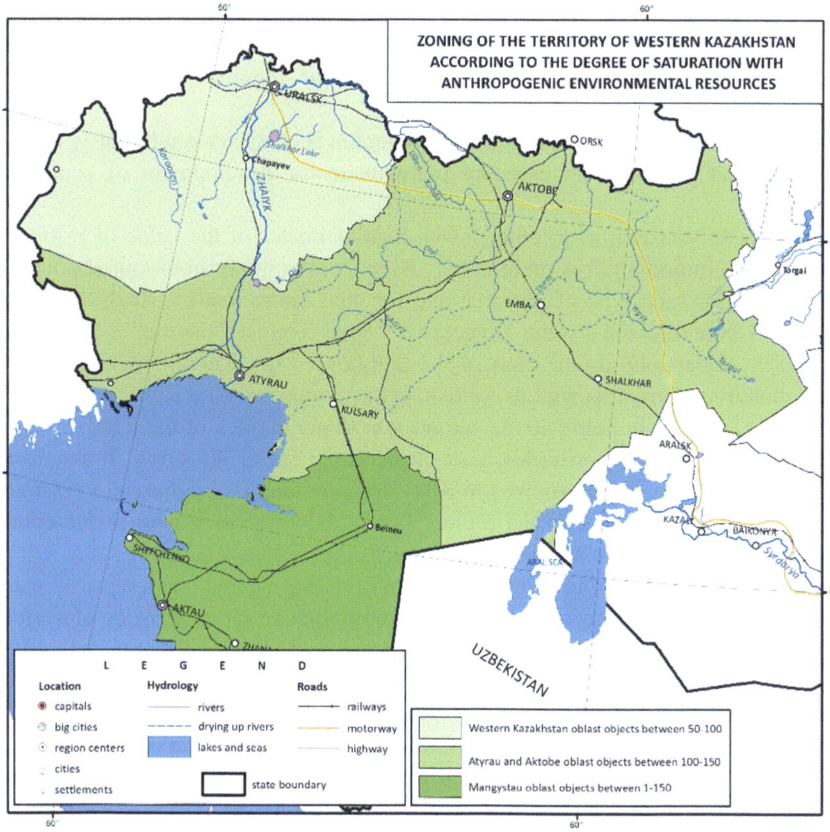

Fig. 5.2 Zoning of the territory of West Kazakhstan in accordance with the degree of satiation of anthropogenic environmental sources (own authors)

One of the area attractions of the fourth group (0-points) is the elevation (terrain) of Mantobe, located in 25 km from Irgiz, where in 1731 years with the participation of Abylkhair (khan of the Younger Zhuz) the "Act of voluntary accession of Kazakhstan to Russia" was signed. In 1981, an obelisk was installed on Mantobe in honour of the 250th anniversary of this historic event (Report on research work 2013; Dmitriev 2010; Geography of domestic tourism of Kazakhstan 2014).

So, the use of anthropogenic recreational resources with the natural resources, in future, can turn the city of Uralsk, Kargaly district of Aktobe region into centres for mass tourism development.

However, a significant number of monuments on history, archaeology and architecture (more than 90%) of the region are in unsatisfactory condition. In this regard, large-scale work is needed to identify, comprehensive study, restoration, tourist and recreational use (for example, by organizing the natural and historical parks) and protection of various monuments. Cognitive tourism resources are important in the education and upbringing of younger generation, in promoting knowledge of nature, economy, and population of Kazakhstan. In the future, resources of this particular type of tourism in the republic can attract foreign tourists.

Thus, the existing tourist potential of the region is characterized by fairly diverse structure and serves as an essential prerequisite for development of ecological tourism.

In tourism research, as in other fields, final estimates of the value of resources involved are important. It is preferable to have them in the form of tourist potential assessment for each type of resource and for their entire totality. Such an assessment can be expressed by the concept of "Total Tourist Potential" (TTP). TTP should be understood as the sum of the addition of the natural, socio-economic recreational resources, as well as tourism infrastructure of a certain territory.

ATP, expressed in points (0–3), allows you to get an idea of the relative, i.e. in comparison with other territories, size of the entire set of resources. It should be noted that the prevailing resources in ATP structure later determine the tourist and recreational specialization of the region. Figure 5.3 presents a map of the cumulative tourist potential in West Kazakhstan.

In general, the ATP is an essential element of managing and planning sustainable tourism development. Accordingly, there are qualitative characteristics of various recreational resources and the level of development on infrastructure of certain areas that are promising for recreation and tourism.

Evaluation of the ATP is also necessary for domestic tourism development, the importance of which will increase in future. Evidence of this can be considered as the published works of Yerdavletov (Yerdavletov and Artemyev 2007; Geography of domestic tourism of Kazakhstan 2014), Shabelnikova, etc.

In conclusion, it should be noted that areas with high and average ATP indicators can be assigned to the territories of prior tourist and recreational development in West Kazakhstan, while areas with low and really low rates are oriented for long-term development.

5.2 Tourist and Recreational Potential of North Kazakhstan

Territory of North Kazakhstan is among the developed regions of Kazakhstan.

The region has a certain range of recreationally significant properties applicable to the organization of summer and winter recreation in different groups of population. It is possible to create centres of year-round recreational, therapeutic and educational recreation; and there are prerequisites for the cultivation of skiing, hiking and boating.

The natural tourist and recreational resources. Special features of tourist and recreational resources in North Kazakhstan include the diversity of climatic conditions and presence of pronounced unfavourable areas. Analysis of mineral water

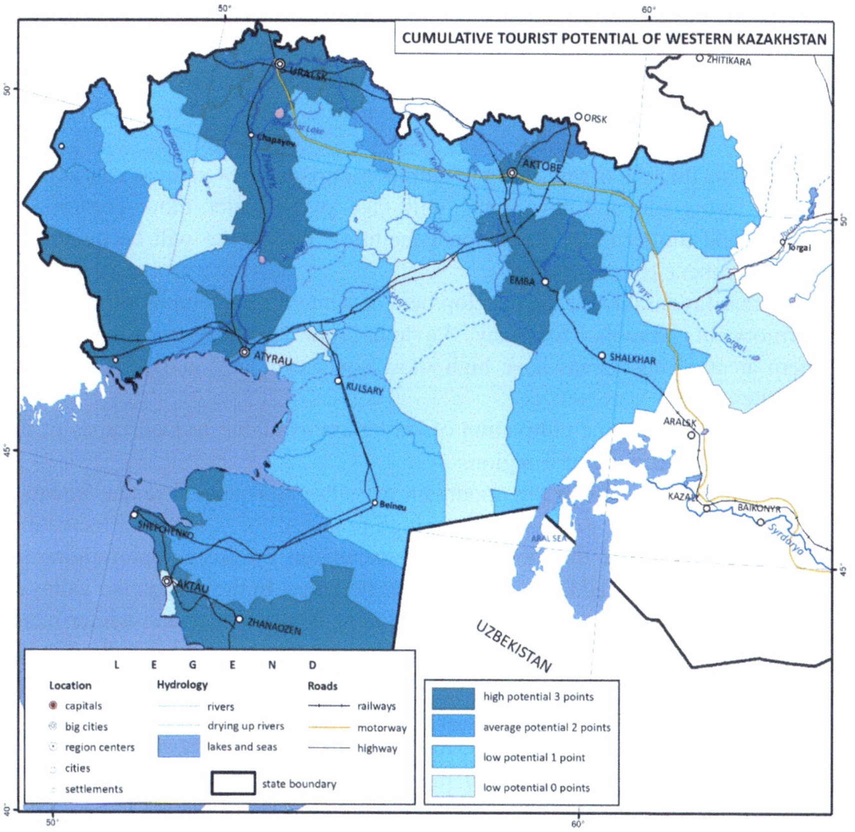

Fig. 5.3 Cumulative tourist potential of West Kazakhstan (authors)

resources shows that in many the natural areas of North Kazakhstan there are sources that can become the basis for creation of therapeutic recreation facilities. Evaluation of therapeutic mud indicates that these resources are practically absent in the south and central part of the region. The highest estimated points received the northern and eastern regions, as well as the floodplain Yertis. They have mineral water sources, and the territory of Kokshetau upland and North Kazakhstan plain has healing mud. Part of the resources on mineral waters and therapeutic mud is successfully used in the climatic and balneological sanatorium-resort institutions of the tourist system in North Kazakhstan (Report on research work 2013; Geography of domestic tourism of Kazakhstan 2014; Titova 2007).

Territory of North Kazakhstan is the most favourable for organizing summer types of tourism. These areas are characterized by the most optimal combination of thermal regime in territories, and temperature of surface waters and vegetation. Among areas with high estimates of the nature concerning vegetation, thermal conditions, watering, i.e. the areas favourable for summer tourism development, and include a large part of the Kazakh hillocky area (Akmola region, Pavlodar region). A group of limited favourable and unfavourable in the degree of attractiveness on the natural landscapes is formed by the southern desert areas of Kostanay region territory. Here it is possible to organize sports and research types of tourism.

It is obvious that the organization of summer tourism types in areas classified as extremely favourable and favourable will take place under more comfortable conditions, and the terms of recreational facilities functioning will be quite long (April–September).

Main areas favourable for the development of winter tourism types are located in the territory of Akmola, Kostanay, Pavlodar and North Kazakhstan regions (northern areas). These areas have high snow cover estimates and persistent negative temperatures.

Good conditions for the cultivation of skiing are available in landscapes of the northern and central parts of a region.

Territories associated with low mountain massifs—Borovoe, Zerenda, Imantau, Bayanaul, Yereymentau, Sandyktau have high potential for year-round recreation. The northeastern part of the territory is characterized by great opportunities for development of summer and selectively winter holidays. In the east, in the valley of Yertis river, there are opportunities for development of summer and winter recreation, although not all of its forms. The natural potential of Turgai hollow is not favourable for recreation and tourism. Selectively areas with moderate potential are found throughout the region.

From the point of recreation view, the most valuable are dry, unwetted areas, which create favourable hygiene and microclimatic conditions. The presence of large wetlands precludes the organization of recreation and tourism.

The above shows that the territory of North Kazakhstan is quite different in the composition of the natural resources used in organization of tourist activities, in their size and combination, as well as in the natural and economic factors that facilitate or impede their development. Forests and pine forests, resort resources (steppe air, mineral waters, and curative mud), scenic areas, surface waters, hunting and fishing grounds make the study area very attractive for tourism and recreation development.

In assessing the natural conditions for recreation, the most important criterion is the diversity of the natural environment, which puts forward landscapes as the main object of recreational assessment.

The best ground for the assessment of tourist and recreation sources and griddling territory in accordance with the degree of attraction of the natural landscapes, in the opinion of most scientists, is a landscape map.

When considering the natural prerequisites (topography, climate, water resources, flora and fauna) and determining the tourist attractiveness of the natural landscapes in the territory of North Kazakhstan, both the descriptive method and point-based assessment method proposed by Milevskaya (1967) classified types of the natural landscapes according to their attractiveness for recreation and tourism. In accordance with this methodology forms of native landscapes were classified pursuant to 4 administrative areas geographically bound to North Kazakhstan. The result of classification on forms of landscapes was the number of points, which evaluated relief, water, forests and extra attraction objects of nature. Taking the scale of value from 1 to 4 points, four districts with different degrees of tourist attractiveness are identified. Applicable system of native conditions and assessment of resources allows comparing various areas and reveal possibilities of geographical environment for mass tourism types development:

1. very favourable areas—4 points (Akmola region);
2. favourable—3 points (Pavlodar and Kostanay regions);
3. unfavourable—2 points (North Kazakhstan region);
4. adverse—1 point (–).

Map names "Zoning the territory of North Kazakhstan according to the degree of the natural landscapes attractiveness" provides more visual information about the areas where the natural recreational resources are concentrated (Fig. 5.4). For its compilation, materials from the Landscape Map of Kazakhstan (Report on research work 2013; Geography of domestic tourism of Kazakhstan 2014; Titova 2007), Maps of the nature reserve fond of Kazakhstan with other cartographic and literary sources were used. The zoning map of North Kazakhstan territory was compiled from extensive field and camera research with the use of existing component maps and literature.

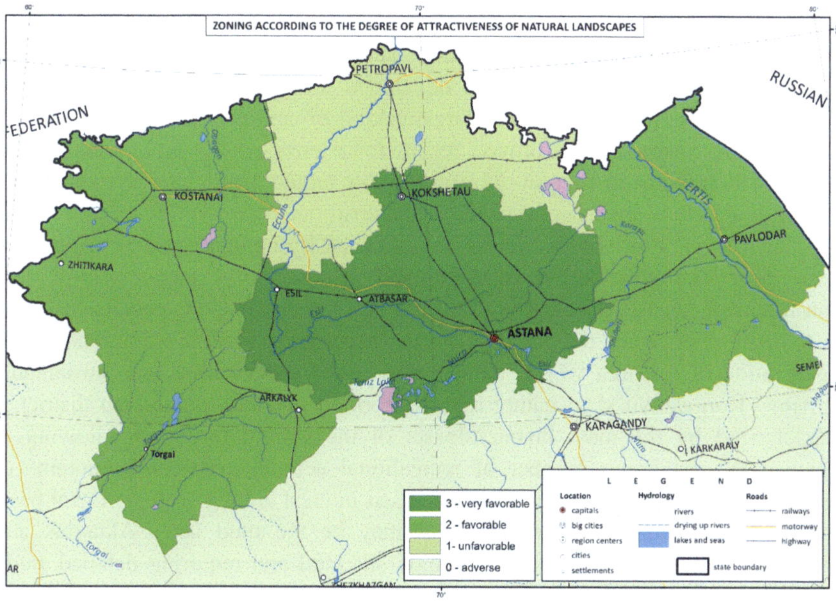

Fig. 5.4 Zoning according to the degree of the natural landscapes attractiveness in North Kazakhstan (authors)

Physical-geographical approach allowed to identify the following main groups of North Kazakhstan landscapes according to their recreational quality and possibilities for organizing various types of recreation and tourism. It is based on landscape zoning.

1. Landscapes with very favourable conditions for a long and all types of short-term rest (3 points). The analysis shows that low-mountain forest landscapes of Kokshetau upland and Yereymentau physico-geographical province are favourable for the organization of summer and winter tourism types.
2. Landscapes with selectively favourable conditions (2 points) for all types of short-term recreation: individual and mass. These include low-mountain and low-slope steppe landscapes of the Kazakh folded country.
3. Landscapes with unfavourable conditions for recreation (1 point) form forest-steppe, dry-steppe low-slope, dry-steppe flat, semi-desert small-slope and semi-desert flat landscapes. These include a number of the Siberian Plain areas, the Atbasar dry-steppe plain province, the Teniz depression. They are characterized by poorly dissected relief, the complete absence of woody vegetation, a few the natural and artificial reservoirs.

Analysis of the natural prerequisites' assessment for tourism and recreation development in the North Kazakhstan and landscape classification according to the degree of attractiveness made it possible to draw some conclusions (Report on research work 2013; Geography of domestic tourism of Kazakhstan 2014; Titova 2007).

The most attractive recreational resources are concentrated in the Korgalzhyn, Shchuchinsk, Yereymentau, Sandyktau, Zerenda districts of Akmola region. Low-mountain forest landscapes are developed recreationally due to favourable climatic and orographic conditions. Organization of activities in Borovoe national parks on the Shchuchinsk-Borovoe resort zone territory and Kokshetau on the territory of Zerenda district in Akmola region turned them into recreation and tourism zones that are in compliance with territorial recreation systems TRS on republican level. Borovoe district is distinguished by unique climatic factors that have a beneficial effect on human health, and air ionization is higher than in other high-mountain areas. Korgalzhyn district is recommended as a specialized centre for tourism, science and the natural history. There are favourable conditions for the organization of research works, here is one of the unique objects of sustainable tourism—Korgalzhyn state nature reserve, formed in 1958. Sandyktau district is characterized by the most favourable set of the natural resources, allowing to organize the maximum number of recreational activities. Falcon mountains of Yereymentau region are a subject of increased interest among tourists. Local basis for the organization of resort treatment may be the mud lakes Aktaylak and Zhaksytuz, containing continental sulphide silt mud and requiring detailed study provided. With appropriate organization of tourism activities, development of infrastructure, construction of new and reconstruction of old roads in the described region areas, it is possible to organize international tourism and form a TRS of international rank. Forests, diverse relief, rich flora and fauna, abundance of lakes, salubrious climate, radon springs, and beauty of landscapes distinguish these landscape zones from others.

1. For the development of certain tourism types, the territories of Kostanay and Pavlodar regions are promising. Low-mountain and low-slope landscapes, the presence of pine and birch logs, band forests, reservoirs, unique nature monuments, sources of mineralized water, mud, clay, having balneological significance, proximity to industrial centres contribute to the organization of recreation areas and tourism of regional and local rank. These areas are recommended as general health because the presence of mineral springs, mud lakes makes it possible to build new health centres here. The most promising for development are radon waters in the north of Akmola region, city of Makinsk.

Balneological value of these waters gives grounds for recommending the construction of cardiological health centre. These areas are unevenly developed in tourism, and favourable the natural recreational resources are used spontaneously and irregularly. But, given the factor that there are a significant number of reserves, nature monuments (both existing and prospective), the territory is favourable for the ecological tourism development.

Rich forest-steppe and lakes of the Akkol region are located in the central part of Akmola region. The little known territory says that tourism is practically undeveloped here. The Akkol and Bulandy districts, considering their advantageous transport and geographical location and the socio-economic development of Stepnogorsk city, have significant tourist potential for the formation of TRS on republican rank.

2. Territory of North Kazakhstan region is the least attractive for mass tourism.

Forest-steppe lowland and dry steppe landscapes offer little conditions for comfortable rest. However, the interest of tourists to exotic, adventure in extreme conditions increases the attractiveness of these landscape zones, and given the increased interest of foreign tourists to steppe safari and their sufficient economic development, the organization of adventure and exotic tourism is possible. It is also advisable to recommend such territories for amateur rest, which do not require large expenditures for development.

Districts with a variety of landscape features that allow planning the development of both educational and sports tourism include individual parts of the Kazakh upland; to zones of possible multidisciplinary development of tourism—the valleys of large rivers (Ishim, Nura) and adjacent territories, to which large group systems of population settling in region (Report on research work 2013; Geography of domestic tourism of Kazakhstan 2014; Titova 2007).

Climatic conditions of a region, with the presence of mineral waters, create prerequisites for balneotherapy facilities, focused on the treatment of digestive organ diseases, organs of the musculoskeletal system and other diseases.

Development of tourism requires a combination of both the natural and man-made recreational resources.

Socio-economic and recreational resources. Zoning and evaluation of area for tourism and recreation purposes demands on complicated factors—nature, economic, physiological, environment, social, architecture, artist, and many others. Recreational zoning methods rely on different parameters as zoning scale and purposes:

- determining tourist and recreational potential of the area for possible study for the development of economy;
- looking for ability of potential rest for population in certain places and its planning;
- placement of specific tourist places in planned rest areas and organizing its planning process in detail.

Relief, waters, vegetation, climate, outstanding objects of nature and culture are evaluated for ecotourism purposes.

It has been studied the experience of different countries in application of different methods of zoning, it can be concluded that the territory zoning is carried out with one or two dominant criteria, which in most cases are the natural and anthropogenic resources.

A comprehensive study and analysis of the anthropogenic recreational resources in North Kazakhstan allows us to conclude how rich the cultural heritage of the great steppe Kazakh people.

As a result of research, the following map was drawn up "Zoning the territory of North Kazakhstan according to the degree of saturation with socio-economic tourist and recreational resources" (Fig. 5.5).

The most studied one, where main significant and attractive objects are shown in the amount of more than 1000, is the North Kazakhstan region—4 points; from 1000 to 900—Kostanay region—3 points; from 900 to 800—Akmola region—2 points; from 800 to 700—Pavlodar region—1 point. Nur-Sultan, Petropavlovsk, Kostanay, Pavlodar, Kokshetau, where monuments of monumental art are significant, stand out among the cities (Report on research work 2013; Geography of domestic tourism of Kazakhstan 2014; Titova 2007).

Analysis of socio-economic conditions and factors on tourism development in North Kazakhstan and regionalization of the area in accordance with the saturation degree with anthropogenic tourist and recreation sources displayed the need to inventory tourist facilities in the region according to their importance, safety and to carry out restoration works of the most significant tourist sites (see Sect. 3.1).

The wider use of cultural and historical sites in sightseeing tourism is hampered by the fact that many of them are not restored, there are no access roads and surrounding territories are not landscaped. With proper organization and preparation of monuments to receive visitors, the expected tourist flow to these sites will be very significant. At the same time, it is very difficult to use cultural and historical sites to disperse them throughout the northern region. Only cities have a significant

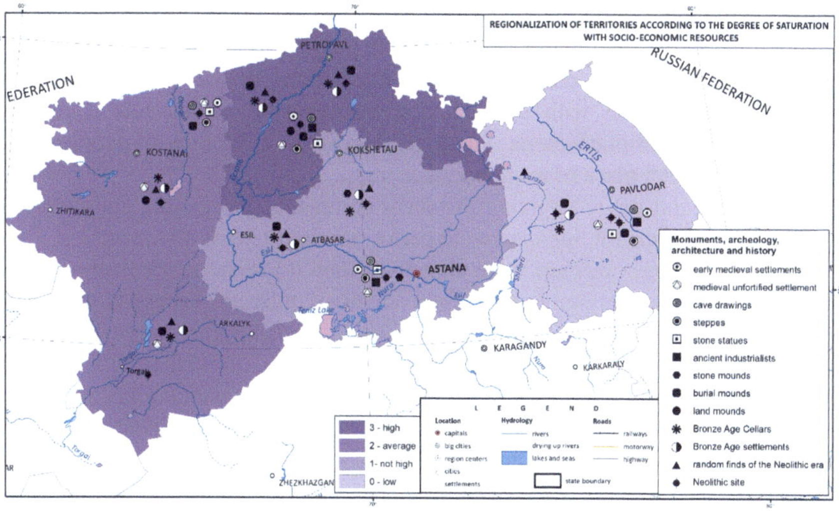

Fig. 5.5 Zoning of territory according to the degree of saturation with socio-economic recreational resources in North Kazakhstan (authors)

number of historical and architectural monuments, moreover differ in the relative concentration of excursion objects.

Zoning grants collecting excursion information on cultural, history and other interesting places in area. Distribution of areas grants local authorities to focus, first, on territory with the highest degree of saturation with anthropogenic tourist and recreational resources; to strengthen measures for further preservation of significant tourist and recreational facilities and their involvement in tourism funds.

Tourism infrastructure. Tourist and recreational economy in the territory of North Kazakhstan was formed intoxicated of local tourist needs and recreational resources and is focused on meeting the demand of population in main types of short and long rest.

Dynamics of tourist flow, which is characterized by a sharp decline in 2000–2005, has a positive trend since 2008, due to a number of market and organizational factors. Annual increase in the tourist flow of domestic tourism was 11%, inbound —7%. Ratio of tourism types indicate the preferential development of domestic tourism, which is a positive factor for regional economy, since it stimulates the creation and modernization of material base and tourism infrastructure on its territory, as well as the production development of goods and services in tourism-related areas (www.stat.gov.kz).

Since 2008, the region has seen a general trend of increasing number on tourism infrastructure facilities by an average of 18% (www.stat.gov.kz). This suggests that entrepreneurs consider promising investment in the development of tourism industry and expansion of tourism services. In Akmola, Pavlodar regions, there are many spa facilities, rehabilitation centres and boarding houses. Sanatoriums are supplied with the advanced medical equipment, allowing highly qualified personnel to work and treat at the highest level.

Analysis of statistical data shows that fluctuations in the volume of tourist services are associated with a qualitative restructuring of the tourist market, determined by social phenomena. Significant decrease in the flow of people going abroad to permanent place of residence, a decrease in the number of people going to shopping tours and the re-targeting of tourist companies to inbound and domestic tourism led to the natural fluctuations in performance of tourist organizations, since outbound tourism services are more expensive than domestic tourism services. In addition, well-known foreign tourist centres do not require any special expenses for their advertising by tourist companies, while advertising for domestic tourism services does not extend beyond a single area. Currently, an increasing number of travel companies are investing in advertising their services, realizing that this is the only way to attract tourists (Report on research work 2013; Geography of domestic tourism of Kazakhstan 2014; Titova 2007).

The territory of North Kazakhstan includes a huge number of facilities that provide places to accommodate tourists and tourist facilities.

According to the survey, many of these objects have seasonal tourist visits and are empty during the downturn of tourists. According to representatives of a region, the main stream of tourists falls on summer and the New Year, when almost all hotels are occupied. This is probably the reason for high prices on living in the region during seasonal time. In competitive countries, on the contrary, during the season, travel agencies offer additional discounts.

As in any other region, tourism is represented by enterprises of communal ownership; private; property of foreign legal entities.

There are several areas of travel agency activity:

- consulting activities;
- provision on services for registration of exit documents;
- provision on tickets for all types of transport;
- organization of inbound tourism and shopping tours;
- organization of recreation for citizens in domestic tourism.

As a rule, most agencies provide a full range of services and only 18.8% of agencies operate in one direction. The following types of tourism are distinguished:

- domestic tourism—includes residents of the country, travelling only within the country;
- outbound tourism—includes residents of a country travelling in another country;
- inbound tourism—includes non-residents travelling to our country.

In 2013, through travel agencies of the Republic of Kazakhstan, 146.9 thousand tourists and sightseers organized their holidays.

Generally, in the northern region of Kazakhstan at present time almost all existing types of tourism are provided—educational, entertaining, ethnic, ecological, and following tours are offered—hunting, rafting, pant treatment, kitting, photo tours, helicopter tours and others. Many travel routes throughout Kazakhstan (www.visitKazakhstan.nur.kz) have been developed for Kazakhstan citizens, as well as foreign tourists. Analysis of these tours showed the presence of huge tourist offer in the territory of North Kazakhstan. This fact is explained by the competent work of tour operators, who form these tours and advertise them on the Internet, which makes it possible for both Kazakhstan and foreign tourists to familiarize themselves with them.

Zoning of the North Kazakhstan territory for purposes of tourism development. A comprehensive analysis of recreational resources, socio-economic factors and current state of tourism sector in North Kazakhstan allows us to estimate the total tourist potential of the resource (TPR).

Development of the country's economy requires that before the development of new resources, before the creation of new tourist complexes, the most accurate quantitative assessment of the whole complex on existing conditions be carried out to minimize any economic miscalculations. Readiness of territories for development may not be correctly evaluated without considering achieved level of its scientific and technical readiness, which is formed by the components of the natural, socio-economic resources and conditions, as well as data on possible ways of their inclusion in economic turnover, obtained as a result of solving these problems. Therefore, before starting the economic development of territory, and to approach the assessment of aggregate tourist resource potential, we must have relevant background information about it (Shabelnikova 1999; Titova 2007).

Aggregate TPR of each potential region is defined as the sum of points, identified by the degree of saturation with the natural, socio-economic recreational resources and provision of tourist infrastructure. The result of this work is the cartogram reflecting territories (areas) with high, medium, low and really low total TPR.

This cartogram shows that the Akmola region territory has the highest potential, having received the maximum number of points—10; average potential, i.e. Pavlodar, North Kazakhstan and Kostanay regions have got 7 points; low potential, which corresponds to 6 points (Fig. 5.6).

Territory of North Kazakhstan has a high potential for developing tourism. This is as a result of the presence of huge amount on recreational resources and favourable conditions of low-mountain forest landscapes.

For the calculation of aggregate TPR in administrative regions of North Kazakhstan, a system of points is also used, calculation of which is carried out in tabular form. According to the evaluation Table 5.1, cartogram named "The aggregate tourist-resource potential of the North Kazakhstan territory" was compiled.

Territory characterized by favourable economic and geographical conditions and a high level of recreational potential includes Akmola region, the best plans for the development of tourism.

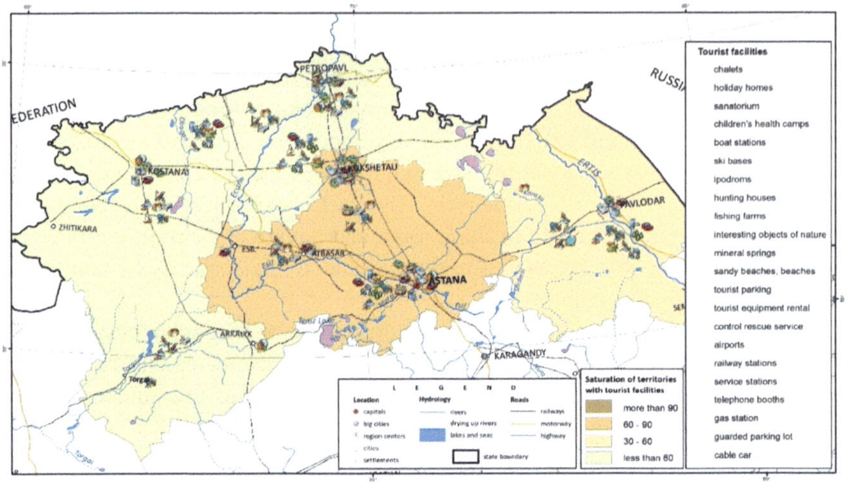

Fig. 5.6 Zoning of the North Kazakhstan territory by total tourist and recreational potential (own authors)

Table 5.1 Calculation of the aggregate tourist-resource potential in North Kazakhstan territory (compiled by M. A. Titova)

Region	Points			Total number
	Saturation degree of the natural recreational resources	Saturation degree of anthropogenic recreational resources	Provision degree of tourist facilities	
Akmola	4	2	4	10
Kostanay	2	3	2	7
Pavlodar	3	3	1	7
North Kazakhstan	2	4	1	7

Areas with relatively favourable conditions and the average level of tourist potential include Pavlodar and Kostanay regions, due to the attractive native landscapes of Bayanaul national park, Nauryzym the natural reserve and numerous native monuments; where the large cities of Kostanay and Pavlodar are characterized by centuries of old history and the presence of significant cultural and historical prerequisites for tourism and recreation development and have sufficiently high level of infrastructure development.

The area characterized by insufficiently favourable conditions and low tourist potential is the territory of North Kazakhstan region considering underdevelopment on tourist infrastructure.

Research has led to the conclusion that the Schuchinsk, Zerenda, Korgalzhyn, Sandyktau districts of Akmola region, Aiyrtau district of North Kazakhstan region, Bayanaul district of Pavlodar region, Nauryzym district of Kostanay region have high and medium total TPR (Report on research work 2013; Geography of domestic tourism of Kazakhstan 2014; Titova 2007).

These are mainly old-developed recreational areas with secure economic and social infrastructure with efficiently studied and developed recreation resources and different native and cultural heritage.

5.3 Tourist and Recreational Potential of Central Kazakhstan

Recreation value of the area is established, first, by properties of the most important native factors for the rest: relief, weather, water, vegetation and animal life. Study of the natural recreational resources in Karaganda region (Central Kazakhstan, only one region) for development of recreation and tourism grants zoning. Regionalization of the area in accordance with the degree of native landscapes attraction gives initial information about the possibilities of resources on compliance with the needs of people for recreation.

For accurate zoning of Karaganda region area current works of different authors also maps, and opportunities of organizing specially protected the natural areas (SPNA) were researched, to create recreation and tourism zones (Nazarchuk 1998).

In Karaganda region there are the natural areas of special status, each of which has undoubted interest as objects of tourism and recreation. Karaganda region ranks third among the regions of Kazakhstan in number of protected areas. From this, it follows that, despite the difficult climatic conditions, the presence of national park, reserves, water bodies of national importance, as well as large industrial centres, increases the tourist potential of the region and is one of the conditions for successful development in territories (Tursinbaeva 2001).

Native life conditions for population are the most important factors which influence or facilitate the development of territory. The degree of auspiciousness for person on totality of native environment and its individual elements strongly impact on resorts network, settlement forms, specifics of the types of tourism and recreation (Lopatina and Nazarevsky 1972). The auspicious characteristics of vital activity among population shall reflect on such aspects of living as recreation and health, psychological and aesthetic perception which concerns the natural conditions that may not be quantified at this stage is most often used by the assessments of expert, traditionally presented in a point scale. An example of the most complete such assessment is the scale developed by Lopatina and Nazarevsky (1972). Analysing the "Map of the natural living conditions assessment among population" in Karaganda region, it can be concluded that in studied region the highest score falls on areas with the most favourable the natural conditions for population life. This is an extensive zone of the Kent-Karkaraly region lowlands—more than 2.5

points; Osakarov, Taldinsk, Kyzylray, Ulytau districts—from 2.0 to 2.5; northwestern and northeastern districts of Balkhash, northern Verkhnesarysu and Srednesarysu regions—from 1.5 to 2.0 points. According to grading scale, the least favourable living conditions of population in semi-desert and desert areas: Moinkum, Zhetykonur, Nizhnesarysu, Karakoiyn, Aral-Karakum are less than 1.5 points.

During evaluation of native conditions for recreation, the most important criteria is variety of native environment, which moves out landscapes as the main object of recreation assessment (Report on research work 2013; Tursinbaeva 2003).

Different levels geosystems may be subject to recreation assessment, but particularly important is the actual landscape, which embodies unity of zone, high-tier and azon, sectoral environmental conditions. Landscape is a native area that contains a regular and connected set of local the natural conditions and all of this concerns human's life. Native complexes are identified as a base of landscape study and may be used to perform such functions which are associated with main types of recreation activities: recreation, cognitive, sports and medical activities.

As a result of the study on the natural recreational resources and integrated assessment of administrative areas in Karaganda region for summer and winter tourism types, a map was drawn up "Zoning of the Karaganda region territory according to attractiveness degree of the natural landscapes", with a detailed classification of the natural landscape types by administrative districts of Karaganda region, where the regionalization methodology of M. Milevskaya was used (Fig. 5.7).

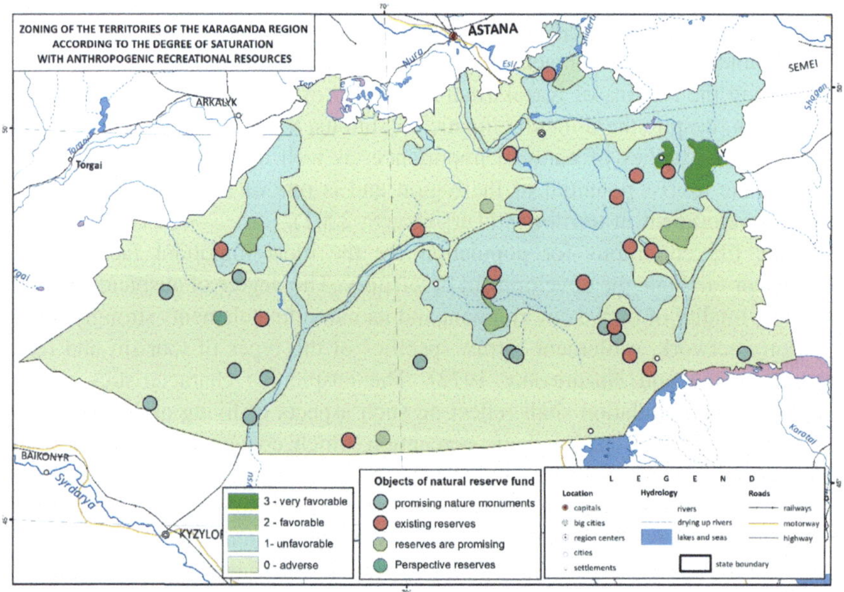

Fig. 5.7 Zoning of the Karaganda region territory by attractiveness degree of saturation with anthropogenic recreational resources (authors)

According to this method, the most attractive the natural landscapes were identified in nine administrative districts of Karaganda region (Report on research work 2013; Tursinbaeva and Abdreeva 2013; Tursinbaeva 2003).

The result of the landscape type's classification, carried out by administrative districts of a region, was the sum of points, which evaluated the relief, water, forests and additional attractive objects of nature. Among administrative districts of a region, the high attractiveness degree of the natural landscapes is distinguished by Karkaraly, with a score of 14; medium—Aktogai, Ulytau—12 points each, low—Abai, Bukharzhirau, Osakarov and Shet—8 points each, low—Nurinsk and Zhanaarka—6 points each.

Considering the physical-geographical approach, the following main groups of landscapes in Karaganda region are distinguished by their recreational quality and opportunities for organizing various types of recreation and tourism (based on landscape zoning conducted by U. A. Kalmenova) (Yerdavletov 2012).

Landscapes with very favourable conditions for long and all types of short-term recreation (4 points), suitable for organizing summer and winter tourism types are low-mountain forest landscapes of Yereymentau—Karkaraly province. In such physiographic areas as Kent—Karkaraly, Kushoky and Taldyn—the share of low-mountain landscape accounts take 80–85% of territory. Low mountains are distinguished by smooth relief, soft outlines of peaks (mountains Kent, Zheltau, Murzhik, Hankashty), covered with pine forests with admixture of birch, aspen, with forbs and feather grass bushes. River valleys, small lakes such as Shaytankol, Pashennoe, Shchuchye give a peculiar attractive appearance. The curative muds of the lake Karasor, and mineral waters of Zhosaly are used for the treatment of holidayers in sanatorium-resort institutions. In summer, the high forest softens the wind force, covers it from the heat, stabilizes fluctuations in air humidity. Lake-mountain-forest landscapes are favourable for organization of sports and recreational tourism and treatment. In winter, the best conditions for organizing sport and tourist ski cycle are created in the north-eastern and central parts of a province.

The landscapes with selectively favourable conditions (3 points) for all types of short-term: individual and mass recreation. These include low-mountain steppe landscapes of Yereymentau—Karkaraly, Kyzyltasskoy and Ulytau provinces.

Landscapes with unfavourable conditions for recreation (2 points) form such dry-steppe low-slope, dry-steppe flat, semi-desert low-slope and semi-desert flat landscapes. These include a number of districts in Tengiz dry-steppe lowland and low-slope provinces, the Ulutau, Sarysu-Tengiz, Kyzyltass semi-desert provinces, which are characterized by poorly divided relief, complete absence of woody vegetation, and small number of the natural and artificial water bodies.

Landscapes with adverse conditions for recreation (1 point)—the desert areas of Shalkar—Nuryn, north—Aral, west—Betpakdala, east—Betpakdala, north—Balkhash provinces. In general, the share of deserts in Central Kazakhstan accounts for about 39.7% of the area under consideration. In recent years, interest in deserts as a resource for tourism has increased (Report on research work 2013; Tursinbaeva and Abdreeva 2013; Tursinbaeva 2003). Deserts have recreational resources for ecology, science and education, and also extreme types of tourism.

Evaluating analysis of the natural basis for the development of tourism and recreation in area and classification of landscape types according to the attractiveness degree led to the following conclusions:

(1) the natural recreational potential of the Karaganda region in conditions of high population demand for rest is insufficient;
(2) the most attractive recreational resources are concentrated in Karkaraly region. Low-mountain forest landscapes are developed respecting recreation due to favourable climatic and orographic conditions and the proximity of Karaganda-Temirtau agglomeration. The organization of national park in Karkaraly forest oasis turned the territory into not only a mass recreation and tourism, which is in compliance with all requirements of territorial and recreational systems (TRS) on the republican scale, but also established good conditions for research. The organization of tourist activities in high level of service guides to the development of infrastructure, road construction in the future, and makes it possibile to organize international tourism and the formation of international trade union among TRSs;
(3) Aktogay, Ulytau and Shet regions are promising for the development of certain tourism types. Low-mountain and low-rise landscapes, the presence of lakes, rivers, reservoirs, proximity of cities and other settlements contribute to the organization of recreation areas on regional and local rank. These areas are unevenly developed in tourist terms, favourable the natural recreational resources are used spontaneously, irregularly (except Balkhash lake region). But, given the factor that there are a significant number of reserves, nature monuments (both existing and prospective), the area may well become a centre for short-term recreation and tourism for residents of Zhezkazgan, Satpayev, and Balkhash. The Balkhash region (north and north-west coasts) and the water area of lake Balkhash have significant the natural potential for the formation of national-level TRSs, considering the socio-economic development of Balkhash industrial hub and advantageous transportation and geographical position;
(4) the most attractive the natural landscapes are located near large settlements, which is important for many industrial enterprises that develop these areas as recreation and tourism.

Socio-economic recreational resources. A comprehensive study and analysis of anthropogenic recreational resources in Karaganda region allows us to conclude how rich the cultural heritage of great steppe Kazakh people. Outside this work, there are many cultural and historical sites that deserve the attention of tourists and have varying degrees of attractiveness. A total of 1981 monuments were identified and registered in the region (Report on research work 2013; Tursinbaeva and Abdreeva 2013; Tursinbaeva 2003).

In Karaganda region, 24 history and culture monuments of state importance are identified, 69 objects are taken under state protection and are considered as objects of regional value, the rest are objects of local importance. Particularly noteworthy are monuments classified as world importance. Such a list establishes and annually updates the UNESCO. The Tamgaly-Tas in Karaganda region also included to the list. Mausoleums of Alash Khan and Zhoshy Khan were restored; materials on the Terekty-Auliye sanctuary were prepared to include to the list of heritage sites of the world.

As a result of research, a map was drawn up "Zoning of Karaganda region territory by saturation degree of anthropogenic recreational resources" (Fig. 5.8).

Data from archaeological map of Kazakhstan, the atlas of Karaganda region, the register of historical and cultural heritage objects of Karaganda region (https://www.google.kz/search?q=Atlases+of+the+Karaganda+region&rlz).

The most studied is Ulytau district, where main significant and attractive objects of show are presented in the amount of more than 400; from 400 to 300—Karkaraly, from 300 to 200—Shet; from 200 to 100—Aktogai, Osakarov, Nuryn; less than 100—Bukharzhyrau, Abay and Zhana arka. Among the cities Karaganda stands out, where monumental art is significant.

The structure of objects is characterized by monuments' predominance:

- archaeology (52.4%);
- architecture (33.6%);
- monumental art (12.4%);
- history (1.6%).

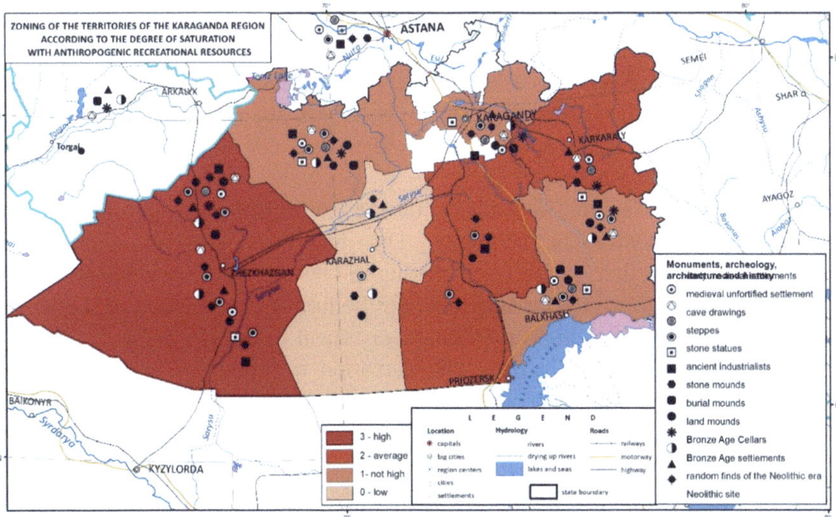

Fig. 5.8 Saturation degree of anthropogenic recreational resources (zoning of Karaganda region territory) (own authors)

Along with the development of tourist infrastructure, it is important not to disturb the historical appearance of the territory. Kvartalnov (2001) recommends the creation of the natural-historical parks, which should help to save the most valuable cultural and historical monuments as integral architectural-landscape and cultural complexes. Moreover, there is a need to apply the ensemble principle and individual design of objects in revival of cultural and historical complexes. Such the natural-historical park was created in the Ulytau region, where archaeological and architectural structures are most compactly located, as well as the picturesque the natural landscapes of the Ulytau and Arganat lowlands (Report on research work 2013).

Analysis of the socio-economic conditions and factors of development in tourism sector of Karaganda region and zoning of this territory according to saturation degree with anthropogenic recreational resources showed the need for:

- inventory of tourist objects in the region according to their importance and safety;
- estimates of anthropogenic tourist and recreational resources as objects of tourism;
- carrying out restoration work of the most significant tourist sites.

The wide use of cultural and historical sites in sightseeing tourism is hampered by the fact that many of them are not restored, there are no access roads and the surrounding territories are not landscaped. Simultaneously, it is difficult using cultural and historical sites to disperse them throughout the region.

It notes that most of the historical and ethnographic monuments of Karaganda region are in disrepair. Therefore, there is a tremendous restoration work, the search for special preservatives, advanced technologies for regeneration of monuments, and training of specialists-restorers. This problem does not only concern the region, but the republic as a whole.

Zoning grants collecting information on the excursion of cultural, history and other places of attention in the region. Distribution of territories allows local authorities focusing, first, on aries with the highest level of anthropogenic recreation sources saturation, to strengthen measures for the further preservation of significant tourist and recreation facilities and their involvement in tourism funds.

Tourism infrastructure. Tourism and recreation facilities in Karaganda region were influenced by local recreational resources and needs, moreover they are focused on meeting the population demand for some main types concerning short and long rest (Report on research work 2013).

Modern tourist infrastructure of Karaganda region is represented by accommodation and catering facilities, transport and communication systems, etc.

Hotel and restaurant business is a leading factor and important component of the tourism economy. According to hotels and restaurant services, business people judge the welfare of the state, for guests it is the face of the country. Tourist flows depend on development level of the accommodation sector.

In 2016, there were 2754 hotel enterprises in Kazakhstan, living in 60,427 rooms.

In Karaganda region, the number of rooms in hotels is 4637 with a lump capacity of 12,441 beds. According to these indicators, the region is inferior to Almaty city, the East Kazakhstan and Almaty regions (Table 5.2). It should be emphasized that in administrative districts there are few hotels with a comprehensive set of services designed to cater for groups and individual tourists, the class and price level of which would be really comparable with the hotel services of international standard. The problem of increasing is that the comfort of rooms remains very relevant.

Zoning of the Central Kazakhstan territory according to aggregate tourist and recreational potential. Comprehensive analysis of recreational resources, socio-economic factors and current state of the tourism sector in Karaganda region makes it possible to estimate the aggregate tourist and resource potential (TRP). Evaluation of the aggregate TRP in territory is important for territorial and regional planning on tourism, for identifying priority development areas in determining the future development of the entire tourism industry and its individual types.

Economy development of any country urgently requires betterment of new resources, creation of new tourist complexes, the most accurate quantitative assessment of the whole complex on existing conditions, and it should be carried out to minimize any economic miscalculations. Readiness of territories for development may not be correctly evaluated without pointing out that the achieved level of its scientific and technical readiness, which is formed by the components of the natural, socio-economic resources and conditions, as well as data on possible ways of their inclusion in economic turnover, obtained as a result of solving these problems. Therefore, before starting the economic development of territory, and to approach the assessment of total tourist resource potential, we must have relevant background information about it (Report on research work 2013; Tursinbaeva and Abdreeva 2013; Tursinbaeva 2003).

Result of this work was the cartogram named "Aggregate tourist and resource potential of Karaganda region", which clearly reflects areas with high, medium, low and really low aggregate TRP (Fig. 5.9).

The first group of territories described by advantageous economic and geographic conditions and importance of recreation possibility comprise Karkaraly district. Karaganda, Temirtau, with the best anticipating development of tourism.

The second group with relatively favourable conditions and with average level of tourist potential includes Aktogay and Ulytau districts. Aktogay district and Balkhash city are highlighted due to the location around Balkhash lake. The Ulytau

Table 5.2 Points evaluation of the tourist and recreational potential in Turkestan region

Regions	Points
Almaty region	High—4 points
Turkestan region	Medium—3 points
Zhambyl region	Low—2 points
Kyzylorda region	Really low—1 point

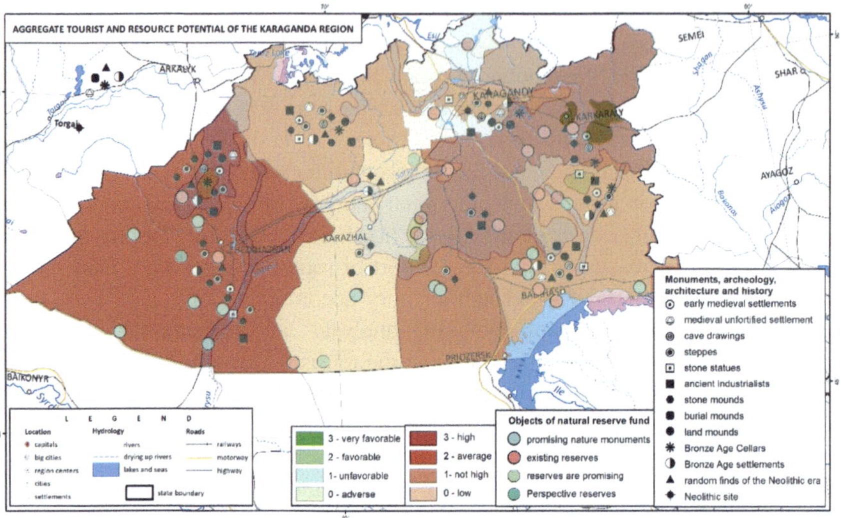

Fig. 5.9 Aggregate tourist and resource potential of the Karaganda region (authors)

district and Zhezkazgan city are considered due to the presence of significant cultural and historical backgrounds, attractive the natural landscapes and fairly high level of infrastructure development.

Among the group characterized by insufficiently favourable conditions and low tourist potential—Osakarov, Bukhar zhyrau, Abay and Shet districts.

And the largest group by territory occupied, but with low recreational potential are Nuryn and Zhana arka districts.

Research has led to the conclusion that territories with high and medium aggregate recreational potential are mainly old-developed recreational areas with settled social and economic infrastructure, owning sufficiently researched and developed recreation sources and diverse native and cultural heritage.

For example, here is the Karkaraly state the natural national park (SNNP), that for Karaganda as urbanized and advanced region is very suitable and perspective territorial form of native conservation, recreational and tourism (Report on research work 2013).

Main objectives of national park: preservation of ecosystems as representative samples on main biotic complexes of a planet; maintaining the ecological diversity of the natural environment; conservation of animal and plant genetic resources; preservation of sites and objects on cultural heritage; preservation of picturesque corners in nature.

The natural resources of a park have high recreational value and are suitable for the development of ecological tourism.

Exploring the total recreational potential and current level of tourism development in Karaganda region, we offer a number of areas in which work can already be started.

(1) Creating your own tourist site of the region in three languages: Kazakh, Russian, English. Site sections may be different (Agency of the Republic of Kazakhstan on Statistics 2010–2012).

- According to the historical places of a region. Show attractive sights: from history, archaeology, architecture and monumental art; museums and exhibitions; existing mosques, temples, holy places for pilgrimage. Make a selection of high-quality photos with objects of show, including beautiful places of nature.

- Modern tourism facilities, where you can talk about where you can have a good time, have fun, eat. Including addresses and features highlight restaurants, bars, saunas, bowling centres, tennis courts, casinos, etc.

- Exotic types of tourism. Here you can offer these types of adventure tourism, such as equestrian, hunting and fishing, including addresses of companies, hunting—fishing farms, offering such services. It is required to be with photos of nature and trophies.

(2) Creation of advertising printed materials in three languages.

- Colour quality booklets about the sights and interesting places of region and cities with brief information about them.
- Brochures "Tourist companies of region" with the name, exact address, and description of the services offered, etc.
- Tourist maps of cities and guides to the area with plotted landmarks (Report on research work 2013; Tursinbaeva and Abdreeva 2013; Tursinbaeva 2003).

(3) Creating an advertising stand of region, for example, "We invite you to travel to Karaganda region". The stand will allow at a decent level to represent the region at exhibitions, fairs, both regional and state levels. Exhibitions and fairs are one of the most effective marketing tools that must be used in full.

Implementation of these events will create the tourist attractiveness of region, its cities and districts, both for domestic tourists and foreigners.

5.4 Tourist and Recreational Potential of East Kazakhstan

The East Kazakhstan economic region consists of the only administrative region—the East Kazakhstan. The total area of East Kazakhstan is 283.2 km^2. Population of East Kazakhstan is more than 1.4 million people. The region is the most sparsely populated among other territories. Its population accounts for approximately 9.1% of the total population in Kazakhstan. Ust-Kamenogorsk is the largest city in the region with a population of 298,100 people.

This region is located in the extreme east of the country—at the junction with China, Mongolia and Russia. Most of the territory consists of high mountain ranges of Ora and Southern Altai, Kalby and Sauyr-Tarbagatai. The height of the mountains reaches 3000–4000 m.

Main rivers are Yertis, Oba and Ulba. The largest lakes within this region are Markakol, Zhaisan, Alakol and Sasykkol. Climate is sharply continental with wide variations in air temperature (Tourism of East Kazakhstan 2013; Bayandinova et al. 2018a, b).

Regarding tourist attractions, special attention should be paid to Katon-Karagay and Kurchum regions, the neighbourhoods of cities named Ridder and Ust-Kamenogorsk, the lakes Alakol and Zaisan, as well as the bank of Bukhtyrma river (Tourism of East Kazakhstan 2013; Bayandinova et al. 2018a, b).

Owning exceptional native sources and original nomadic people culture, there is a huge untapped opportunity for the development of tourism in international and regional markets. Tourist potential of native recreation sources and historical and cultural heritage grants region to harmoniously integrate into international tourism market and reach intensive development of tourism in state. This will provide steady growth in labour and salary of population, by stimulating development of tourism-related industries and development in investing to national economy (Aktymbaeva and Zhaksygalieva 2010; Artemyev et al. 2014).

On the territory there are 12 especially protected native zones, with 1,784,536 ha, and out of which 50,811 ha occupied by water, 641,258 ha belong to forest.

One of the most attractive resources for foreign tourists is Altay mountains.

According to the method of M. Milevskaya (see Sect. 5.2.2), the classification on types of the natural landscapes was carried out for 15 administrative districts geographically tied to the East Kazakhstan. The result of the classification on types of landscapes was the points sum that were used to evaluate the landscapes type, nature monuments, reserves, reservoirs, national parks, mineral water fields, commercial species of the region and other the natural objects that have tourist appeal.

Taking the scale of values from 1 to 4 points, four districts with different degrees of tourist attractiveness are identified. Applied system of assessing the natural conditions and sources grants comparing different areas and revealing possibilities of geographic environment for the development of mass tourism:

1. very favourable—4 points (Katon-Karagay district);
2. favourable—3 points (Zyryanovsk, Kokpekty, Kurchum, Tarbagatai districts);
3. unfavourable—2 points (Abai, Ayagoz, Beskaragai, Borodulikha, Glubokov, Zharma, Zaisan, Ulan, Urzhar, Shemonaikha districts);
4. adverse—1 point (–).

The map “Zoning of the East Kazakhstan territory according to the degree of the natural landscapes attractiveness” (Fig. 5.10) provides more visual information about the areas where the natural recreational resources are concentrated.

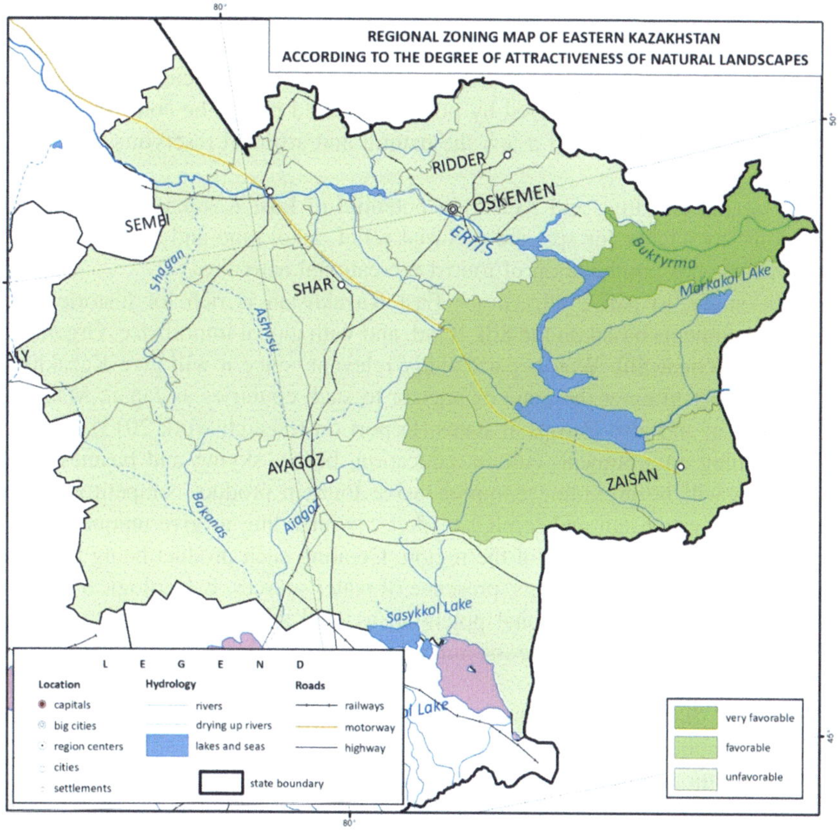

Fig. 5.10 Zoning of the East Kazakhstan territory by the attractiveness degree of the natural landscapes (authors)

Physical-geographical approach allowed identifying the following main groups of landscapes in East Kazakhstan according to their recreational quality and opportunities for organizing various types of recreation and tourism. The basis for landscape zoning (Fig. 5.10):

1. Landscapes with very favourable conditions for a long and all types of short-term rest (3 points). Analysis shows that for the organization of summer and winter tourism types, low-mountain forest landscapes of dark coniferous taiga and larch forests of Kazakhstan Altai and Saur, passes of the Marble Mountain, Azutau ridge, Markakol basin, etc. are favourable.
2. Landscapes with selectively favourable conditions (2 points) for all types of short-term recreation: individual and mass. It includes low-mountain and low-slope steppe landscapes.

3. Landscapes with unfavourable conditions for rest (1 point) form forest-steppe, dry-steppe low-slope, dry-steppe lowland, semi-desert small-slope and semi-desert flat landscapes. These include a number of areas such as Zaisan desert. They are characterized by poorly dissected relief, the complete absence of woody vegetation, and a few the natural and artificial reservoirs.

Key details of native and recreation potential of East Kazakhstan are climatic conditions favourable for spa and pictorial relief, lake, pine and forests, seasonal comfort for types combination of tourist-recreational activities.

Additionally, to native attractions, East Kazakhstan is rich for historical and culture monuments based on the Silk Road, and with global importance. Organizing tours on the Great Silk Way is particularly relevant, since it will give Kazakhstan the opportunity to enter the zone of interest to such countries as Japan, Malaysia, China, Korea, and also European states (Report on research work 2013).

Developing environment, culture, education, beach, skying and business tourism, which will be providing comprehensive tourism product competitive in the Republic of Kazakhstan and foreign markets, is intending to give unique thing to the existing tourism potential of the region. Creating such product being based on advantageous climate conditions, presence of water sources, balneological sources, a rich landscape and recreational potential, as well as an original historical and cultural heritage, the presence of unique museum complexes of international importance.

There are monuments of history and culture (627 historical, architectural and cultural monuments, including 15 which have republican significance) and historical-cultural museums-reserves on the territory of East Kazakhstan which form the core of cultural and educational tourism (Yerdavletov and Artemyev 2008).

Comprehensive study and analysis of the anthropogenic recreational resources in East Kazakhstan allows us to conclude how rich the cultural heritage of great steppe Kazakh people.

As a result of research, a map was drawn up "Zoning of the East Kazakhstan territory according to the degree of saturation with socio-economic tourist and recreational resources" (Fig. 5.11).

From Fig. 5.11 it follows that the main significant and attractive objects in amount of more than 1000 are in Borodulikha district—4 points; from 150 to 200—Abai, Ayagoz, Zharma, Zaisan, Zyryanovsk, Katon-Karagai, Kurchum, Tarbagatai, Ulan, Urzhar districts—3 points; from 100 to 150—Glubokovsk, Shemonaikha districts—2 points; from 0 to 100—Beskaragai district—1 point.

The cities such as Ust-Kamenogorsk, Semey, Zyryanovsk are distinguished, where monumental art is significant.

Analysis of the socio-economic conditions and factors of tourism development in East Kazakhstan and zoning of territory according to the degree of saturation with anthropogenic tourist and recreational resources showed the need to inventory existing tourist facilities in the region for their importance, preservation and restoration of the most important tourist sites.

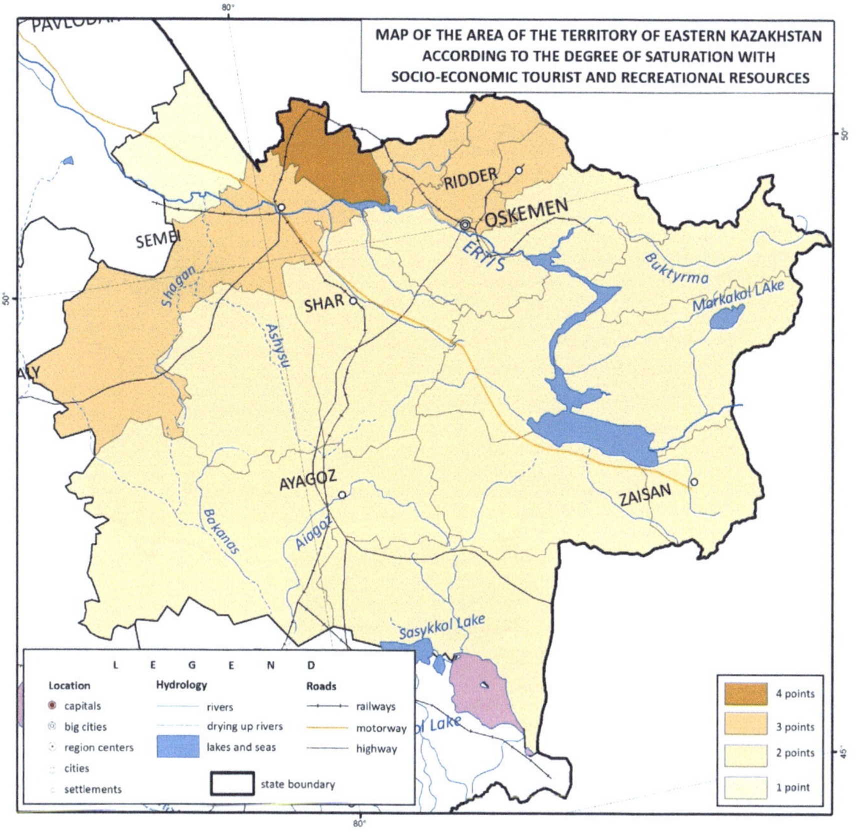

Fig. 5.11 Zoning of the East Kazakhstan territory according to saturation degree of socio-economic tourist and recreational resources (authors)

Tourism industry in East Kazakhstan is recognized at the state level as one of the major sectors of the economy. Thus, in executing provisions of industrial and innovative development in the country's economy, the main role is on the cluster system, in particular the tourist cluster. East Kazakhstan is distinguished as a region in the formation of tourist cluster with specialization of mountain tourism in environmental areas. Current trends in developing this industry are such that tourists, who researched well the most well-known world resorts, is applied to those countries where the tourism sector is just developing. And starting from this position, attraction of Kazakhstan is growing.

Because of comprehensive analysis on the natural and socio-economic recreational resources, map of the total tourist and recreational potential has been compiled (Fig. 5.12).

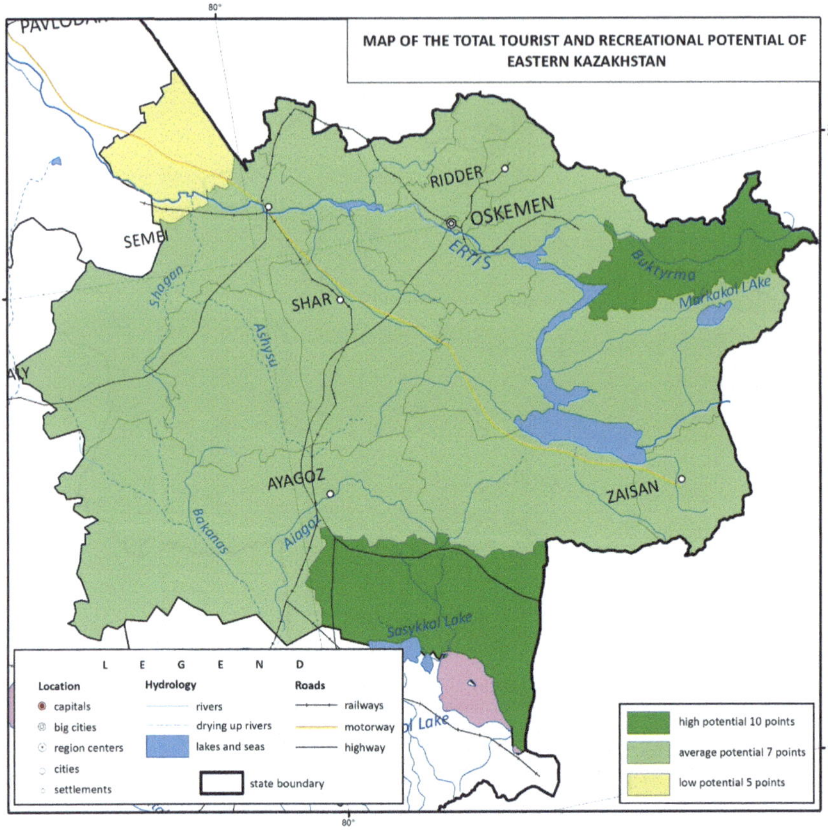

Fig. 5.12 Aggregate tourist and recreational potential of the East Kazakhstan (authors)

This cartogram shows that the highest potential has the territory of Katon-Karagai, Urzhar districts, which received the maximum number of points—10; and average potential, i.e. all other districts of a region has 7 points, except Beskaragai district with low potential—6 points.

As can be seen from figure, the northern and eastern parts of a region are favourable areas for the development of recreational activities.

In general, the territory of East Kazakhstan has rather high potential for tourism development. This is due to the fact of presence of large number on recreational resources and favourable conditions of low-mountain forest landscapes.

Recreational potential allows assessing the possibility and feasibility of developing certain areas of interest for recreational activities, allows to preserve the uniqueness and diversity of the natural landscapes.

Today the area is identified by high rates of economic development; investment attraction; advantageous geopolitic and geographic locations; transportation accessibility; industry of entertainment; advantageous climate conditions; various historic and culture heritage; centuries-old hospitality tradition; presence of education institutions that coach professionals of the tourism industry.

Types of tourism which have potential for development and are recommended for the East Kazakhstan territory:

1. In connection with planned construction of the Olympic base facilities, prospects for sports, entertainment and festival tourism are increasing.
2. Due to historical events of the last century there are great possibilities for developing ethnic and nostalgic tourism.
3. Presences of historical and archaeological objects that are currently not used, allows you to actively develop educational tourism.
4. There are all necessary the natural and infrastructural conditions for the revival of active leisure activities—hiking and mountain tourism. They are almost unlimited resources for the development of climbing.
5. Unlimited opportunities for the development of ecotourism and agritourism.

As part of ecotourism, you can select educational tours, tours in various ecosystems, photo hunting, etc.

When organizing agritourism, the tasks of creating special "tourist villages", the formation of "rural tours" with accommodation and meals in farming families are resolved.

The business tourism is very promising here.

Sustainable development of the natural and recreational systems, as an experimental model of "sustainable" territorial education, must comply with the principles and recommendations of the "Sustainable development concept of the RK", and also strategic goals of the "Program on forced industrial and innovative development of the RK for 2015–2019."

5.5 Tourist and Recreational Potential of Turkestan Region

The natural recreational resources. Turkestan region occupies a vast territory from the Zhongar gate in the east to the Aral Sea in the west and from the desert plateau Betpak-Dala and lake Balkhash in the north to the southern border of the republic, including the northern part of Kyzylkum Desert, the Tien Shan spurs and Zhongar Alatau ranges. Southern Kazakhstan includes Almaty, Zhambyl and Kyzylorda region, Turkestan region This is the region with the highest density of population in the country. There are the largest cities of Kazakstan—Almaty and Shymkent. Main rivers are Syrdarya, Ile, Aksu, Shu, Karatal, Lepsy. The largest lakes: Balkhash, Sasykkol, Aral Sea, Alakol. Climate of Turkestan region is mild, and excellent for tourism and recreation, skiing, climbing and hunting (see Sects. 2.1, 2.2 and 3.1).

Tourist and recreation sources unite native and anthropogenic (culture) landscape, made to manufacturability degree intoxicated of social requirements and the opportunity of their direct usage in tourist and excursion services (Yerdavletov et al. 2014).

Picturesque landscapes of the Almaty region, representing the greatest interest from the tourism point of view, there are mountain systems of Ile and Zhetysu Alatau, Ketmen, which belong to the Northern Tien-Shan.

The Almaty region is also called Zhetysu, that is, Semirechye, which means seven rivers flowing through the territory of the region.

To obtain more visual information on the areas of concentration concerning the natural tourist and recreational resources of Turkestan region areas with sufficiently large potential for the tourism development were identified. These include areas of the south-eastern and eastern parts of Almaty region and Turkestan region. They are characterized by picturesque terrain, forests, the presence of water bodies, mineral and mud wells, and are characterized by sufficient long climatic states favourable to people, and also have many native objects that are interesting for the development of tourism.

Using the experience of various countries, we made an attempt to regionalize the Turkestan region territory according to the saturation degree with recreational resources (the natural and man-made) and the provision of tourist services.

The zoning based on analysis of recreational resources covers not only developed, but also promising territories in terms of tourism development and was carried out within the administrative and territorial boundaries of Turkestan region.

Since the main objects of tourism are landscape elements, then, first, zoning of the natural landscapes was carried out according to the tourist attractiveness degree. The basis of this zoning lay to the landscape map of Kazakhstan.

Study of this map has contributed to the selection of landscapes that have different degrees of favourableness for tourism development: very favourable, favourable and not very favourable.

Landscapes characterized by the presence of picturesque mountain relief, water arteries with significant forests were considered to be very favourable.

Landscapes that are characterized by the presence of two main the natural factors and absence of one of them (for example, there is no forest or water space) are classified as favourable.

Landscape should be considered unfavourable, where of the three main factors there is only one the natural factor contributing to the tourism development.

Zoning of the natural landscapes according to the degree of their attractiveness was carried out in accordance of the zoning method by M. Milevskaya, and it classified the types of the natural landscapes in accordance to their attractiveness for recreation and tourism. The classification result was the sum of points, which were estimated as relief, water, and forests. Territory of Turkestan region was estimated by the reduced system based on landscape map for 4 regions. Each area has a certain number of points. The higher recreational qualities are dominated by Almaty and South Kazakhstan regions, which have landscapes of middle mountains and foothills. They are located in favourable the natural conditions of the highlands.

Unfavourable landscapes are more typical to Kyzylorda region. This is because most of the area is located in desert zone.

As a result of the conducted research, territory was assessed according to the saturation degree with the natural tourist and recreational resources, where following were revealed: the number of points over 151 is in Almaty region; 101–150—in Turkestan region; 51–100—in Zhambyl region; 0–50—in Kyzylorda region. Consequently, the region with maximum number of points has very favourable the natural conditions for tourism development. It belongs to high category (over 151); area where the number of points ranges from 101 to 150—to the average; region, with a rating from 51 to 100 points—to low category; areas with a minimum score of 0 to 50 fall into the low category (Fig. 5.13).

Anthropogenic tourist and recreational resources. Anthropogenic (cultural-historical) recreational resources are one of the major factors determining the state of tourist attractiveness in a particular territory. The areas long inhabited by man keep the history, traditions, culture. Cultural and historical recreational resources are prerequisite for the organization of cultural and educational types of recreational activities; on this basis it optimizes recreational activities in general (Report on research work 2013; Yerdavletov et al. 2014).

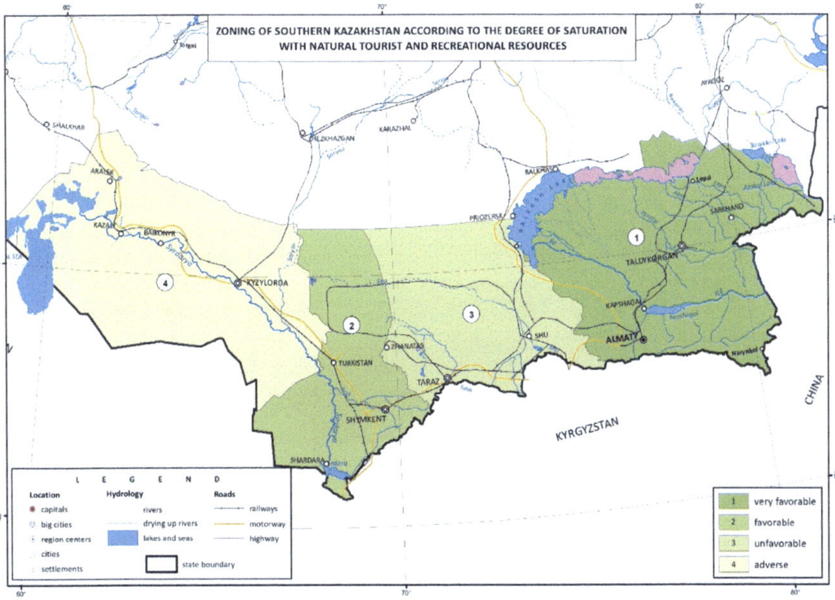

Fig. 5.13 Zoning of Turkestan region according to the saturation degree the natural tourist and recreational resources (authors)

Among cultural and historical monuments, the leading role is played by historical—cultural monuments, which are the most attractive. Typology is based on the informational essence of cultural-historical object: uniqueness, typicality among objects of this kind, cognitive and educational value, visual appeal.

In Turkestan region there is a significant number of monuments establishing the history of Semirechye from ancient times to present day. Most are archaeological gravestones, including settlements, burial yards and grounds, petroglyphs, ancient irrigation-structures, ancient mines and sculptures. They are mainly in the foothill-zone with favourable native conditions, which was the main reason for developing this territory by people since ancient times (see Sects. 2.2 and 3.1).

Analysing the anthropogenic tourist and recreational resources of Turkestan region, it is possible to assess the territory according to saturation degree of cultural and historical tourist and recreational resources. The Almaty region received over 400 points; from 301 to 400—Turkestan region; from 201 to 300—Zhambyl; less than 200—Kyzylorda region. Consequently, the Almaty region, characterized by large number of points, and which is the most favourable for tourism development and is rated as high category territory; the Turkestan region, within which there are between 300 and 400 points, has generally favourable conditions and belongs to territories of the middle category; Zhambyl region, with a score between 201 and 300 points is assigned as low category; Kyzylorda region belongs to the territories with low category (Fig. 5.14).

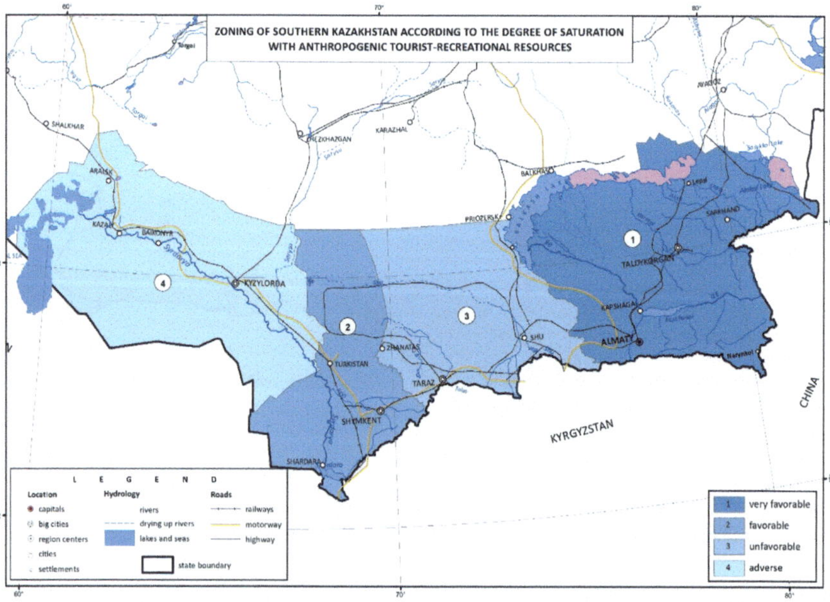

Fig. 5.14 Zoning of Turkestan region according to the saturation degree of anthropogenic tourist and recreational resources (authors)

Tourism infrastructure. One of the basic concepts of tourism geography is the concept of "tourist attractiveness" on object locality. The attractiveness of a particular area is determined not only by the presence of favourable the natural conditions or historical sites, but also by the presence of developed tourist economy (tourism infrastructure).

Tourism industry is a certain complex of various sectors, the creation and development of which is stimulated and developed by the arrival and the stay of tourists. This economy consists of mutually complementary industries that can satisfy the needs of tourists only as a complex (Report on research work 2013; Yerdavletov et al. 2014).

According to the generally accepted structure of tourist economy, institutions and facilities designed to serve tourists are divided into four bases:

(1) overnight stay (accommodation);
(2) communications;
(3) food;
(4) accompanying.

The base on overnight stay consists of tourist-hotel complexes, tourist centres, sanatorium and health complexes, children's health institutions, etc.

Because of mineral springs area, the network of resorts was created. The most well-known are Alma-Arasan, Kargaly, Koktem, Ak-qaiyn and others located in the picturesque gorges of Ile Alatau, and Kapal-Arasan in Zhetysu Alatau, Saryagash in the Turkestan region. Environs of Almaty city have the highest degree of saturation with medical and recreational facilities, where there are most of all the sanatorium and recreational complexes, individual sanatoriums and children's recreational facilities. This is due to the presence of favourable the natural and climatic conditions and resources, as well as the proximity to them of a city with a million people population.

As a result of tourism infrastructure analysis in Turkestan region, the territory was assessed according to the availability degree of tourist services: the number of points over 80 has Almaty region; from 71 to 80—Turkestan region; 61 to 70—Kyzylorda; less than 60—Zhambyl region. Therefore, the most suitable for tourism development is the Almaty region, which has more extensive network of tourism infrastructure. It belongs to the high category; the region, which has from 71 to 80 points and, accordingly, belongs to middle category, and has favourable conditions for tourism development; area within which the number of points varies from 61 to 70 has low category; and area of less than 60 points falls into the low category (Fig. 5.15).

The calculation of total tourist recreational potential was made according to assessment provision in Turkestan region with native recreation, tourist and recreation sources, anthropogenical and tourist facilities.

For convenience of calculating the total tourist-resource potential, a score was assigned to each category in ascending order from 1 to 4.

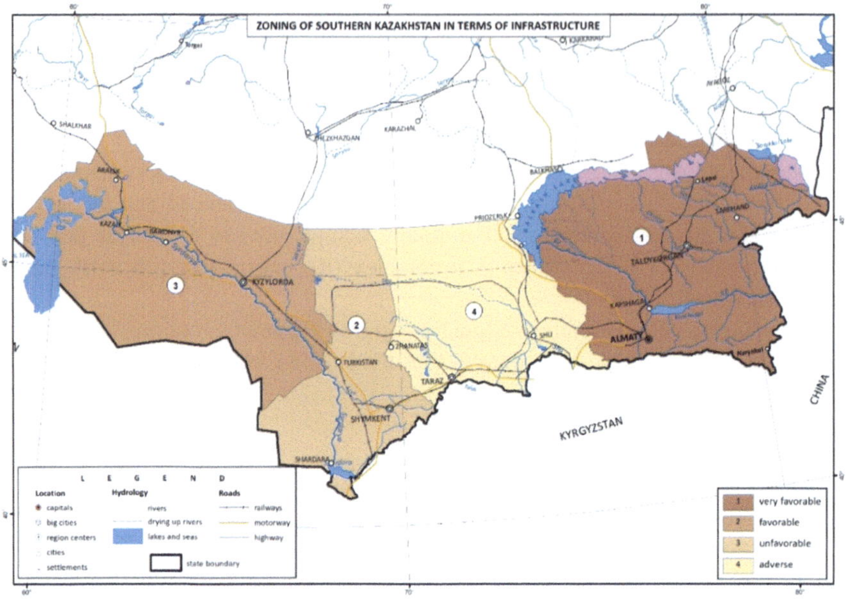

Fig. 5.15 Zoning of Turkestan region by provision degree of infrastructure objects (authors)

As a rule, any quantitative assessment of one or several recreational resources should be reduced to the determination of their total potential. Total, or cumulative, tourist and resource potential (TRP) is a comprehensive aggregate indicator characterizing the possibility of developing tourism in Turkestan region territory.

The aggregate TRP of each district is defined as the sum of points, determined by saturation degree of the natural, man-made recreational resources and the provision of tourist infrastructure.

Calculation of the total TRP is listed in Table 5.2.

Result of this work was the evaluation table and compilation of cartograms. Cartograms clearly show regions with a great perspective for developing tourism (Table 5.3).

In assessing the aggregate TRP, a summary scale was built, also ranked by intervals, and the "Aggregate tourism and resource potential of tourism development in Turkestan region" was drawn (Fig. 5.4), which shows that Almaty region has high potential (12), maximum number of points—4; the average potential (9) is characteristic to Turkestan region, which corresponds to the average score—3; low potential (5) for the Zhambyl region—2 points; and low potential (4) has Kyzylorda region—1 point (Fig. 5.16).

As a result of general assessment of recreational resources, it was revealed that the Almaty region has the best conditions for tourism development (Report on research work 2013; Yerdavletov et al. 2014).

Table 5.3 Calculation of total tourism and resource probability for developing tourism in South Kazakhstan region

Region	Points			Total
	In accordance with the degree of saturation of native recreation sources	In accordance with the degree of saturation of anthropogenic recreation sources	According to the infrastructure provision	
Almaty region	4	4	4	12
Turkestan region	3	3	3	9
Zhambyl region	2	2	1	5
Kyzylorda region	1	1	2	4

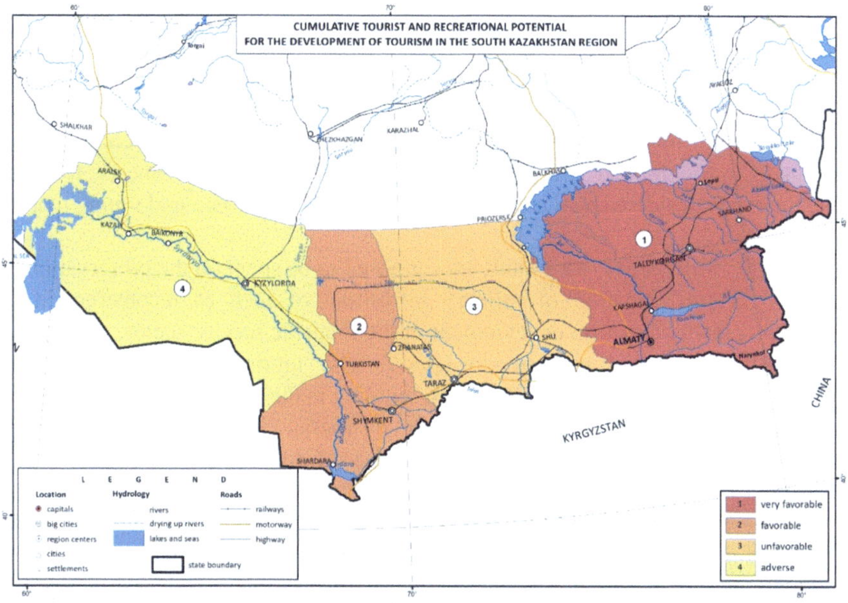

Fig. 5.16 Cumulative tourist and recreational potential of tourism development in the South Kazakhstan region (authors)

Thus, it can be said that in general Turkestan region has a great potential for tourism development. This is due not only to the presence of significant amount on tourist and recreational resources, but also the favourable the natural conditions of foothill and mountain areas.

5.6 Comprehensive Tourism and Recreation Potential of the Republic Kazakhstan

Analysing and summarizing all the data for regions of Kazakhstan, the total tourist and recreational potential of the country. The aggregate tourist and recreational potential of tourism industry development in the Republic of Kazakhstan shows that the north-western regions, i.e., have the greatest security on tourist-resource potential (4 points) as following: the northern part of Aktobe region, 3 points—the northern and northeastern regions, i.e. Akmola region and the eastern part of East Kazakhstan region (Tables 5.4, 5.5, 5.6 and Fig. 5.17).

Conclusion on the 4th section. Kazakhstan, the largest landlocked country, has chosen water-rich areas near rivers and lakes as the basis for its historical development. The territory of the state is 2724.9 thousand km^2, the population is 18 million inhabitants, while a significant part of the territory is not used in economic activities. Historical situation has contributed to the preservation of flora and fauna in remote regions of Kazakhstan. In this regard, in Kazakhstan there are a significant number of protected areas and localities—10 reserves and 12 national parks. In addition, there are 2 UNESCO World Heritage Sites in the south of Kazakhstan, and one in the north.

Today, the state sets tasks of creating a competitive tourist industry to ensure stable growth of incomes and employment in the hospitality sector by enlarging the volume of inflowing and internal tourism, considering that it is domestic tourism, which can be considered as the catalyst for international tourism and tourist market development.

In these priority areas, state funds are supposed to be spent on the preservation of historical centres, creation of primary infrastructure such as reconstruction of highways, railway network, development of the roadside service sector, and others.

Construction of small and medium-sized hotels, "tourist villages", bringing the level of existing hotels service to 3–4 stars, building in the Karkaraly mountain forest oasis, Ulytau, Balkhash lake and Balkhash city tourist cluster facilities, creating a chain of motels and campgrounds on highways of Almaty-Balkhash-Nur-Sultan, Karaganda-Karkaraly, Karaganda-Zhezkazgan-Ulytau—all this is supposed to be done at the expense of private investors.

Most of the sights (both existing and potential) are associated with nature, but at the same time in south of Kazakhstan there are historical and cultural objects, primarily associated with the Silk Road.

Considering the above, there is a plan within the framework of Silk Road program to submit on UNESCO applications for inclusion in the World Heritage List of 10 additional sites in Kazakhstan.

In general, the use of tourist and recreational resources is directed primarily at domestic tourism, which is less demanding in terms of quality and diversity of the services offered. Often the problem is infrastructure that prevents more intensive use of a particular tourist object. The access to objects in most cases is difficult, for example, Sharyn canyon.

Table 5.4 Tourist attractiveness of the natural resources of Kazakhstan

Regions and districts of Kazakhstan	Landscape types			Monuments of nature	Reserves	Reservoirs	Naticnal parks	Dendroparks	Botanical gardens	Zoos	Reservoirs of national importance	Mineral water deposits	Game animals			Others	Total
	3 points	2 points	1 point										Animals	Птицы	Рыбы		
1	2	3	4	5	6	7	8	9	10	11	12	13	14	15	16	17	18
West Kazakhstan	–	–	20	–	–	3	–	–	–	–	–	2	–	–	–	6	31
Aktobe	–	–	39	–	–	1	–	–	–	–	–	–	4	1	–	6	51
Atyrau	–	–	13	–	–	1	–	–	–	–	1	1	–	2	3	3	24
Mangistau	–	–	7	–	1	2	–	–	1	–	1	3	6	3	3	8	35
West Kazakhstan region	**–**	**–**	**79**	**–**	**1**	**7**	**–**	**–**	**1**	**–**	**2**	**6**	**10**	**6**	**6**	**23**	**141**
Kostanay	–	4	25	–	1	3	–	–	1	–	–	3	2	5	–	5	49
North Kazakhstan	1	4	4	12	–	4	1	–	–	–	3	4	–	–	–	3	36
Akmola	2	3	20	8	1	5	3	–	–	–	–	1	1	2	30	5	81
Pavlodar	–	1	16	1	–	2	1	–	–	–	–	–	1	–	–	9	31
North Kazakhstan region	**3**	**12**	**65**	**21**	**2**	**14**	**5**	**–**	**1**	**–**	**3**	**8**	**4**	**7**	**30**	**22**	**197**
Karagandy	–	6	30	–	1	11	1	–	1	1	1	5	5	4	3	12	81
Central Kazakhstan region	**–**	**6**	**30**	**–**	**1**	**11**	**1**	**–**	**1**	**1**	**1**	**5**	**5**	**4**	**3**	**12**	**81**
East Kazakhstan	10	47	42	1	2	4	1	–	1	–	2	1	8	4	–	19	142
East Kazakhstan region	**10**	**47**	**42**	**1**	**2**	**4**	**1**	**–**	**1**	**–**	**2**	**1**	**8**	**4**	**–**	**19**	**142**
Kyzylorda	–	–	12	–	1	2	–	–	–	–	1	–	4	–	3	–	23

(continued)

Table 5.4 (continued)

Regions and districts of Kazakhstan	Landscape types			Monuments of nature	Reserves	Reservoirs	National parks	Dendroparks	Botanical gardens	Zoos	Reservoirs of national importance	Mineral water deposits	Game animals			Others	Total
	3 points	2 points	1 point										Animals	Птицы	Рыбы		
1	2	3	4	5	6	7	8	9	10	11	12	13	14	15	16	17	18
Turkestan region	2	7	7	–	2	5	1	1	–	1	–	5	2	–	–	3	36
Zhambyl	–	4	19	–	1	3	–	–	–	–	1	2	5	1	2	5	43
Almaty	2	24	14	4	2	8	4	1	2	1	4	–	9	8	4	7	94
Turkestan region	**4**	**35**	**52**	**4**	**6**	**18**	**5**	**2**	**2**	**2**	**6**	**7**	**20**	**9**	**9**	**15**	**196**
Republic of Kazakhstan	**17**	**100**	**268**	**26**	**12**	**54**	**12**	**2**	**6**	**3**	**14**	**27**	**47**	**30**	**48**	**91**	**757**

Bold Total score for regions and RK

Table 5.5 Saturation of the Kazakhstan territory on socio-economic recreational resources

Regions and districts of Kazakhstan	Archaeological monuments	Architectural monuments	Monuments	Memorials	Museums and exhibitions	Petroglyphs	Theaters, palaces of culture	Circuses	Sport facilities	Temples and mosques	Interesting objects of national economy			
											Industry	Agriculture	Transportation	Social complex
1	2	3	4	5	6	7	8	9	10	11	12	13	14	15
West Kazakhstan	18	1	13	7	21	–	9	–	1	4	10	2	5	1
Aktobe	5	3	–	3	47	–	8	–	1	8	10	3	–	3
Atyrau	11	–	–	3	37	–	4	–	2	1	1	3	2	2
Mangystau	10	16	3	3	14	–	6	–	3	4	10	2	7	1
West Kazakhstan region	**44**	**20**	**16**	**16**	**119**	**–**	**27**	**–**	**7**	**17**	**31**	**10**	**14**	**7**
Kostanay	25	3	3	7	28	–	8	–	3	2	11	3	4	2
North Kazakhstan	12	4	2	–	28	–	9	–	1	15	11	3	3	1
Akmola	43	1	5	5	54	–	18	1	2	1	11	3	3	4
Pavlodar	23	1	2	1	31	–	6	–	2	–	12	3	5	1
North Kazakhstan region	**103**	**9**	**12**	**13**	**141**	**–**	**41**	**1**	**8**	**18**	**45**	**12**	**15**	**8**
Karaganda	126	2	2	2	55	2	14	1	2	13	12	3	4	2
Central Kazakhstan region	**126**	**2**	**2**	**2**	**55**	**2**	**14**	**1**	**2**	**13**	**12**	**3**	**4**	**2**
East Kazakhstan	30	4	8	1	30	1	6	–	2	5	12	3	4	1
East Kazakhstan region	**30**	**4**	**8**	**1**	**30**	**1**	**6**	**–**	**2**	**5**	**12**	**3**	**4**	**1**
Kyzylorda	20	3	4	1	22	1	3	–	1	11	11	3	4	1

(continued)

Table 5.5 (continued)

Regions and districts of Kazakhstan	Archaeological monuments	Architectural monuments	Monuments	Memorials	Museums and exhibitions	Petroglyphs	Theaters, palaces of culture	Circuses	Sport facilities	Temples and mosques	Interesting objects of national economy			
											Industry	Agriculture	Transportation	Social complex
1	2	3	4	5	6	7	8	9	10	11	12	13	14	15
Turkestan region	41	1	15	1	57	2	19	1	1	16	12	3	4	1
Zhambyl	28	4	4	1	24	4	6	–	1	10	12	3	4	2
Almaty	63	26	19	5	92	14	43	1	5	–	12	3	5	5
Turkestan region	**152**	**34**	**42**	**8**	**195**	**21**	**71**	**2**	**8**	**37**	**47**	**12**	**17**	**9**
Republic of Kazakhstan	**455**	**69**	**80**	**40**	**540**	**24**	**159**	**4**	**27**	**90**	**147**	**40**	**54**	**27**

Regions and districts of Kazakhstan	Interesting events or phenomena			Ethno centres, museums	Centres of handicraft	Others	Total
	Ethnographic nature	Political character	Economic character				
1	16	17	18	19	20	21	22
West Kazakhstan	–	1	–	7	1	95	196
Aktobe	5	1	–	3	–	66	166
Atyrau	–	1	1	2	–	29	99
Mangystau	–	1	–	2	–	32	114
West Kazakhstan region	**5**	**4**	**1**	**14**	**1**	**222**	**575**
Kostanay	–	1	–	–	–	91	191
North Kazakhstan	7	1	–	3	–	55	155
Akmola	1	5	1	9	–	129	296
Pavlodar	–	1	–	–	–	77	165
North Kazakhstan region	**8**	**8**	**1**	**12**	**–**	**352**	**807**
Karaganda	1	1	1	1	–	88	332
Central Kazakhstan region	**1**	**1**	**1**	**1**	**–**	**88**	**332**
East Kazakhstan	4	1	–	–	–	99	211
East Kazakhstan region	**4**	**1**	**–**	**–**	**–**	**99**	**211**

(continued)

Table 5.5 (continued)

Regions and districts of Kazakhstan	Interesting events or phenomena			Ethno centres, museums	Centres of handicraft	Others	Total
	Ethnographic nature	Political character	Economic character				
1	16	17	18	19	20	21	22
Kyzylorda	1	1	–	1	–	71	159
Turkestan region	1	1	–	1	1	105	283
Zhambyl	–	1	–	1	–	61	166
Almaty	2	6	1	2	–	117	421
Turkestan region	**4**	**9**	**1**	**5**	**1**	**354**	**1029**
Republic of Kazakhstan	**22**	**23**	**4**	**32**	**2**	**1115**	**2954**

Bold Total score for regions and RK

Table 5.6 Objects of tourist economy in Kazakhstan

Regions and districts of Kazakhstan	Travel agencies and tour operators	Tourist centres	Hotels	Guest and hunting houses	Sanatoriums, rest houses	Ski resorts	Health camps	Airports	Railway stations	Bus stations	Marine stations, wharves	Service stations
1	2	3	4	5	6	7	8	9	10	11	12	13
West Kazakhstan	37	0	44			0	4	1	6	4	2	3
Aktobe	41	0	47			0	8	1	3	1	0	8
Atyrau	41	0	43			0	4	2	10	5	1	5
Mangistau	20	1	53			0	6	1	10	7	3	6
West Kazakhstan region	**139**	**1**	**187**			**0**	**22**	**5**	**29**	**17**	**6**	**22**
Kostanay	62	1	81			0	15	1	19	8	0	5
North Kazakhstan	20	1	53			0	19	1	13	1	0	5
Akmola	323	15	361			0	22	2	38	7	0	12
Pavlodar	57	0	59			0	27	1	3	2	0	3
North Kazakhstan	**462**	**17**	**554**			**0**	**83**	**5**	**73**	**18**	**0**	**25**
Karaganda	189	5	150			0	13	3	7	3	0	11
Central Kazakhstan region	**189**	**5**	**150**			**0**	**13**	**3**	**7**	**3**	**0**	**11**
East Kazakhstan	64	4	265			1	22	2	29	5	0	15
East Kazakhstan region	**64**	**4**	**265**			**1**	**22**	**2**	**29**	**5**	**0**	**15**
Kyzylorda	11	0	58			0	9	1	1	1	0	7
Turkestan region	101	1	104			0	17	1	4	13	0	8
Zhambyl	30	0	62			0	9	1	23	3	0	6
Almaty	709	9	274			4	6	2	18	8	2	12
Turkestan region	**851**	**10**	**498**			**4**	**41**	**5**	**46**	**24**	**2**	**33**
Republic of Kazakhstan	**1705**	**37**	**1654**			**5**	**181**	**20**	**184**	**67**	**8**	**106**

(continued)

Table 5.6 (continued)

Regions and districts of Kazakhstan	Gas stations	Catering points	Objects of entertainment					Supermarkets	Customs	Water sources	Total
			Water parks	Technoparks	Night clubs, discos	Casinos	Bowling				
1	14	15	16	17	18	19	20	21	22	23	24
Regions and districts of Kazakhstan	Gas stations	Catering points	Objects of entertainment					Supermarkets	Customs	Water sources	Total
			Water parks	Technoparks	Night clubs, discos	Casinos	Bowling				
1	14	15	16	17	18	19	20	21	22	23	24
West Kazakhstan	9	65	0	1	3	0	2	10	11	3	205
Aktobe	8	132	0	0	3	0	5	13	14	4	288
Atyrau	5	155	0	0	5	0	10	3	8	1	298
Mangistau	6	83	1	0	6	0	11	2	10	1	227
West Kazakhstan region	**28**	**435**	**1**	**1**	**17**	**0**	**28**	**28**	**43**	**9**	**1018**
Kostanay	5	67	4	1	6	0	8	8	20	2	313
North Kazakhstan	5	20	1	0	3	0	4	12	13	3	174
Akmola	13	24	2	0	32	6	93	46	0	4	1000
Pavlodar	4	66	0	0	20	0	6	7	11	3	269
North Kazakhstan	**27**	**177**	**7**	**1**	**61**	**6**	**111**	**73**	**44**	**12**	**1756**
Karaganda	12	133	2	1	13	0	34	28	0	4	608
Central Kazakhstan region	**12**	**133**	**2**	**1**	**13**	**0**	**34**	**28**	**0**	**4**	**608**
East Kazakhstan	17	125	2	1	14	0	9	14	11	5	605
East Kazakhstan region	**17**	**125**	**2**	**1**	**14**	**0**	**9**	**14**	**11**	**5**	**605**
Kyzylorda	7	25	1	0	2	0	4	2	2	3	134
Turkestan region	8	93	2	1	13	0	13	11	15	3	408
Zhambyl	6	55	2	0	5	0	3	4	6	1	216
Almaty	21	250	5	4	58	20	65	66	5	9	1547
Turkestan region	**42**	**423**	**10**	**5**	**76**	**20**	**85**	**82**	**28**	**16**	**2305**
Republic of Kazakhstan	**133**	**1293**	**22**	**9**	**181**	**26**	**267**	**225**	**126**	**46**	**6292**

Bold Total score for regions and RK

Many of the objects listed on the official website visitkazakhstan.kz do not represent objects of interest (for example, "monuments of the city"). At the level of regional tourism development plans, it is necessary to conduct a thorough analysis of sights to sort them according to the criterion of attractiveness for tourists arriving from far and near abroad, and for tourists inside the country.

The influence of tourism on the economy of the country is demonstrated by indicators of cash-flows in and out of the country, which are originated from tourism imports and exports. Most of the tourist services income from export comes out of arrival of foreign tourists, and from tickets sale for local transport and other in host country services. In-country tourist imports are summed up of money disbursed by its citizens on foreign trips, transportation expenses and other in-host country services, as well as benefits to tourism industry foreign investors.

Difference between the income of the country from the tourist services export and importing tourist services cost is balance of payments of tourists, which may be negative (deficit) as well as positive.

In world practice, tourism is classified mainly as inbound, outbound and domestic, and inbound tourism is considered to be the main generator of foreign exchange earnings. The structure of tourism industry in Republic of Kazakhstan is similar to the world one and includes 3 components—exit, internal, and entrance.

Comprehensive study and assessment of the natural-geographical conditions and resources, as well as other regional characteristics of the territory are important conditions for recreational economy development. Based on a combination of factors, it is possible to determine which types of tourism are most efficiently developed in a given territory.

Problem in the country is the lack of access roads to many tourist sites and lack of facilities in surrounding areas. The proper organization and preparation of monuments to receive visitors can help the expected tourist flow to these sites will be very significant. Along with this, it is very difficult to use cultural and historical sites to disperse them throughout the republic. Only cities, which have a significant number of historical and architectural monuments, differ in relative concentration of sightseeing objects.

Realizing existing prospective is possible only under the systematic and comprehensive conditions attitude solving problems that hinder development of tourism.

Thus, although native monuments of Kazakhstan are not only under most in the world's address to the world, and often beat them, use of them for tourism aims is still tough due to low development of tourism infrastructure in general. To fundamentally change the current situation, the following steps are necessary:

- to solve the problem of using nature monuments for tourist purposes at state level by adopting new legislative acts and resolutions on their protection regime and to strengthen control over their observance;
- to strengthen funding for study and preservation of the natural monuments, to interest in this international organizations such as UNESCO, UNICEF, etc.;

- it is necessary to develop a plan for prospective use on each of nature monuments and culture of the Republic of Kazakhstan: new opportunities for its protection, maintaining proper cleanliness and order, advertising, information and cartographic support (http://www.pm.kzru; http://www.kta.kz; www.pm.kzru).

As shown by the results of tourism potential research, Kazakhstan has great capacity and possibility for the future development of ecological tourism. This statement is based on uniqueness of native conditions and landscapes in Kazakhstan, large number of historic monuments, native, cultural, ethnic legacy of people who dwelled the territory in different historical periods.

Currently, development of ecological tourism is researched mostly in protected regions, and projects shall be developed for the development of ecological tourism in rural areas.

In regions of Kazakhstan, the most relevant investment projects are being worked out, for which support is provided by the Ministry and promotion to pull investments for the tourism infrastructure development. So, the following investment projects were approved and recommended on further implementation:

- in the Akmola—ethnographic complex "Sheberauly" with exhibition centre "Palace of masters", recreation centre "Sharzhum", recreation area "Kunbay Sulu"; state historical and cultural nature reserve "Tamgaly", tourist and ethnographic complex "Talkhyz";
- in the Atyrau—construction of recreation centre in Sarytogai rural area of the Makhambet district;
- in the East Kazakhstan—the health-tourist complex "Katon-Karagay", health-improving complex "ANA", tourist cluster "Altai Alps", tourist route "Altai—the golden mountains";
- in Zhambyl region—the tourist centre "Tau Samaly".

Realization of these projects will guide to important diversification of tourism industry, and will increase its innovative and infrastructural components, moreover will create the prerequisites for successful development of non-primary sectors.

The most efficient to manage protected regions is zoning. Usually, zoning is included to the projects of functioning of protected regions, and monitoring their progress is available to management. Dwelling should minimize any breaches or invasions to the wildlife, nature and usage of this tourism may be organized. As an example, tourism may be unacceptable in reproducing areas or assailable the natural environment, where any human encroachment affects the nature. Reasonable and viable tourism depends on biological zone features, native species susceptibility, as well as economical and human resources are required for organizing protected regions, native parks and range of the services given to tourists. At the same time, specialists must necessarily assess the ability of visited territory to withstand tourism stresses. This considers the number of visitors, compile statistics of visits, and, if necessary, stop tourists flow timely.

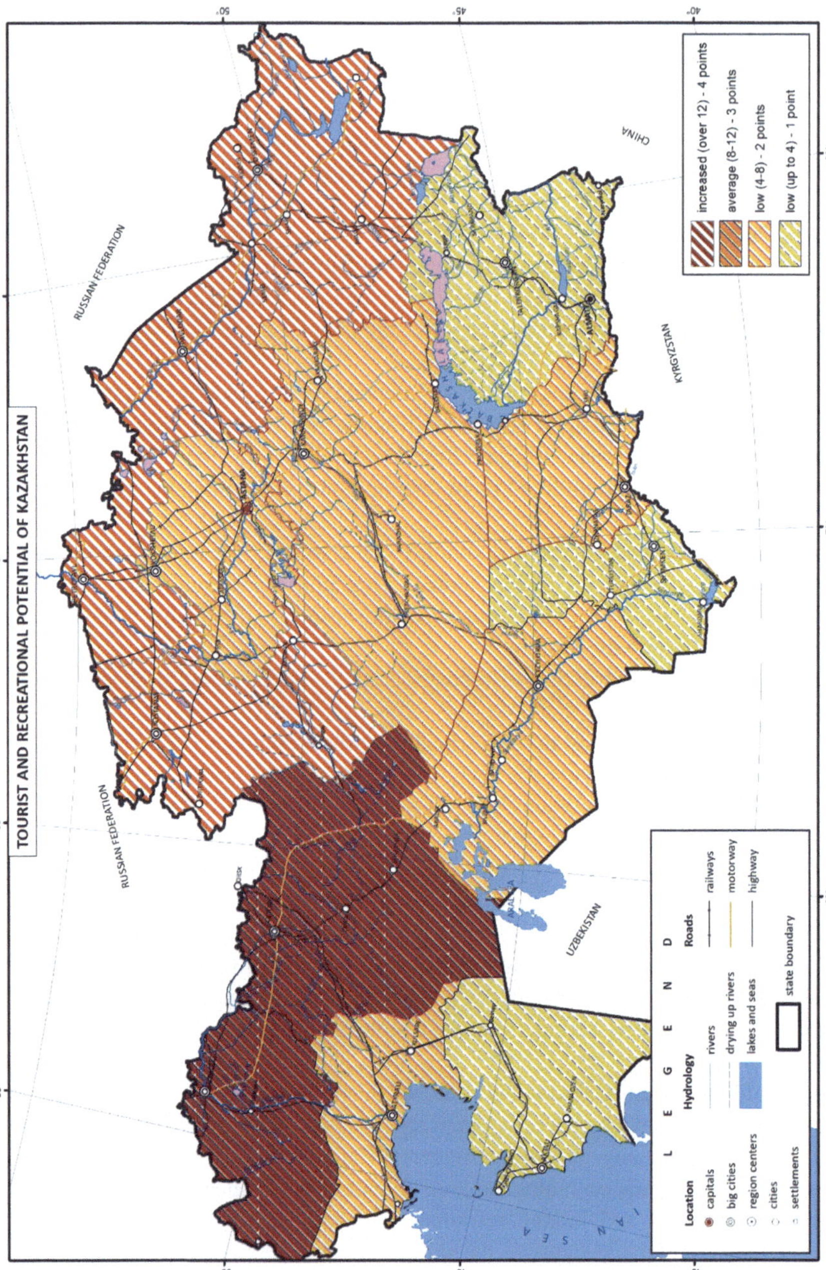

Fig. 5.17 Tourism and recreation potential of Kazakhstan (own authors)

Consequently, enormous potential of ecotourism should be implemented with correlation management of government. For ecotourism organization and finance benefits to conservation on wildlife, shall be in correlation with pricing policies which reflect on real services expenses for tourists and cost on operations of native objects.

Ecotourism is an important source of income from nature conservation. There are the following sources of income:

- payment for environmental management in admission form, payment for the use of transport and accommodation, hunting trophies, etc.;
- rent of individuals for the provision of services in food form, housing, transport, guide;
- percentage from the sale of products to tourists visiting this territory;
- taxation of goods and services.

A number of authors propose to consider ecological tourism as a potential means ensuring the conservation of biological resources.

Regardless of what, developing ecotourism shall pursue purposes of biodiversity preservation, with a certain profit. The following conditions will influence it:

- areas of tourist-visited shall be characterized by the existence of interesting wildlife components, achievable for viewing, meaning that they must be exclusive or able for the attraction of visitors;
- being in a charge of protected areas shall have legal authorities for development and implementation of sustainable tourism in the SPNAs;
- there should be implementation of efficient pricing policy which will take into consideration gathering fees, solvency of people and amount of which depends on using ecotourism expenses (Report on research work 2013).

Finally, all funds which are received shall be allocated primarily to specially protected the natural areas required and all biodiversity conservation in the country.

The practice shows that when the local population is involved in ecotourism projects, such violations as, for example, indiscriminate cutting down of trees or poaching are significantly reduced.

At the same time, this engagement should not only be in the form of alms and charitable assistance, or even the food supply to schools, hospitals and social services funded from tourist income. It is important that the local population itself takes an active part in ecotourism activities.

In Kazakhstan, the negative effects of ecotourism are not so problematic, but rather we can talk about the consequences of just tourism in some vulnerable habitats of wild animals. However, they are sufficiently tangible. For example, consequences of visiting the nesting sites of some rare birds (flamingos, sickle birds, and others) that is very sensitive to anxiety factor, which are unpredictable. Amateur ornithologists, to photograph them on their nests, are quite close to the birds. Currently, many species of animals are in dysfunctional state due to the violation of their habitat by humans. All this makes today think about ways of developing ecological tourism in the country, considering the protection and rational use of the natural environment.

Thus, from all the above, we can draw the following conclusions:

1. In conditions concerning transition of Kazakhstan to sustainable development, reorientation is necessary, reassessment of the priorities on functioning of the natural-economic systems to obtain the maximum economic and environmental effect from ecological tourism development.
2. Ecotourism makes a major contribution to nature conservation through the support of protected areas, which play a vast role in maintaining the ecological balance of the planet.
3. In sustainable development concept of the Republic of Kazakhstan, strategic plans in field of environmental protection indicate the development in future on environmentally friendly types of tourism, such as ecological tourism.
4. Today, as part of the state order for national parks, master plans have been developed that also allow optimal determination on planned locations for main tourist facilities in national parks considering environmental requirements and the natural features. Every year the number of visitors to specially protected native regions of Kazakhstan increases.

References

Agency of the Republic of Kazakhstan on Statistics (2000–2012) Tourism. The main indicators for 2000–2012. http://www.newskaz.kz

Aktymbaeva AS, Zhaksygalieva ZR (2010) Regionalnye aspecty ustoichivogo razvitiya turizma v Vostochno-Kazakhstanskoi oblasti (Regional aspects of sustainable tourism development in the East Kazakhstan region). In: Proceedings of the IX interregional scientific and practical conference "Opportunities for tourism development in the Siberian region and adjacent territories", Tomsk, pp 72–75

Arheologicheskaya karta Kazakhstana (Archaeological map of Kazakhstan). https://www.google.kz/search?q=Archaeological+Map+of+Kazakhstan&rlz=1C1GGRV_enKZ751KZ751&tbm=isch&tbo=u&source=univ&sa=X&ved

Artemyev AM, Aktymbaeva AS, Yerdavletov SR (2014) Rekreasionnoe zonirovanie Vostochno-Kazakhstanskoi oblasti i opredelenie konseptualnyh napravlenii razvitiya turizma (Recreational zoning of the East Kazakhstan region and determination of conceptual areas for tourism development). Bull d'Eurotalent-Fidjip 61–66

Atlas Karagandinskoi oblasti (Atlases of the Karaganda region). https://www.google.kz/search?q=Atlases+of+the+Karaganda+region&rlz

Bayandinova S, Mamutov Zh, Issanova G (2018a) Man-made ecology of East Kazakhstan. Springer, Singapore, p 144

Bayandinova S, Mamutov Z, Issanova G (2018b) The natural factors of formation and development of geosystems in East Kazakhstan. In: Environmental science and engineering (subseries: environmental science), pp 1–30

Dmitriev MN (2010) Economika turisticheskogo rynka. Uchebnik (Economy of the tourist market. Textbook), 2nd edn. Reworked and additional. Unity-Dana, Moscow, p 311

Geographiya vnutrennego tyrizma Kazakhstana: nauchnaya monographiya (Geography of domestic tourism of Kazakhstan: scientific publication) (2014) Edited by Erdavletov SR. Kazakh University, Almaty, p 300

Government on-line. Strategic plans of ministries in Kazakhstan until 2020. http://www.pm.kz

Kazakhstan Tourist Association (KTA). Reports on the work done by the Kazakhstan Tourist Association (KTA) (electronic resource). http://www.kta.kz

Kvartalnov VA (2001) Mezhdunarodnyi turizm (Foreign tourism). Finance and Statistics, Moscow, p 312

Landshaphnaya karta Kazakh SSR (Landscape map of the Kazakh SSR) (1979) Republic of Kazakhstan, p 1

Lopatina EB, Nazarevsky OR (1972) Osenka prirodnyh uslovii zhizni naseleniya (Assessment on the natural conditions of population life). USSR Academy of Sciences, Institute of Geography, Moscow, p 148

Milevskaya MI (1967) Klassifikasiya turistskih territorii (Classification of tourist areas), 3rd edn. Zemlya, Moscow, p 230

Ministry of industry and new technologies of the Republic of Kazakhstan. Reports on the work done (electronic resource). http://www.pm.kz

Musabaev T, Muldagaliyeva K (2012) Konsepsiya Konseptualnye podhody razvitiya turizma na territorii Respubliki Kazakhstan (Conceptual approaches of development tourism on territory of the Republic of Kazakhstan). REAL CORP, p 320

Nazarchuk MK (1998) Karta zapovednogo fonda Respubliki Kazakhstan (Map of the natural reserve fund in Republic of Kazakhstan). LLP Ecoproject, Almaty, p 1

Ofisialnaya web-stranisa statisticheskih dannyh Raspubliki Kazakhstan (Official website of government statistical data of the Republic of Kazakhstan). www.stat.gov.kz

Ofisialnaya web-stranisa Vizit Kazakhstan (Official website of Kazakhstan tourism attractive places). www.visitKazakhstan.nur.kz

Otchet o nauchno-issledovatelskoi rabote: Razrabotka Konsepsii regionalnogo razvitiya e territorialnoi orzanizasii vnutrennego turizma Respubliki Kazakhstan (promezhutochnyi) (Report on research work: concept development of regional development and territorial organization on internal tourism of the Republic of Kazakhstan (interim)) (2013) Ispolniteli (project executors): Yerdavletov SR, Aktymbaeva AS, Iskakova KA, Aliyeva Zh et al. Almaty, p 181. № Gos registrasiya (State registration) 0112PK02459

Preobrazhenskiy VS (1975) Theoreticheskie osnovy rekreasionnoi geographii (Theoretical foundations of recreational geography). Science, Moscow, p 223

Shabelnikova SA (1999) Metodika kompleksnoi ozenki recreazionnyh resursov dlya razvitiya turizma e otdyha: avtoreferat dissertazii na soiskanie uchenoi stepeni kandidata geographicheskih nauk: 25.00.24 (The method of integrated assessment on recreational resources for development of tourism and recreation: abstract of the dissertation for the degree of candidate of geographical sciences: 25.00.24). Almaty, p 23

Titova MA (2007) Problemy i perspektivy razvitiya i territorialnoi organizasii turizma Akmolinskoi oblasti: avtoreferat dissertasii na soiskanie uchenoi stepeni kandidata geographicheskih nauk: 25.00.24 (Problems and prospects of development and the territorial organization of tourism of Akmola region: abstract of the dissertation for the degree of candidate of geographical sciences: 25.00.24). al-Farabi Kazakh National University, Almaty, p 16

Tourism of East Kazakhstan (2013) Statistical collection. Ust-Kamenogorsk, p 430

Tursinbaeva KS (2001) Karkaralinskii nasionalnyi park i ego rol v razvitii turizma Karagandinskoi oblasti (Karkaraly national park and its role in tourism development of Karaganda region). In: Materials of the international conference "The role of tourism in sustainable development of the Republic of Kazakhstan". Kazakh University, Almaty, pp 117–123

Tursinbaeva KS (2003) Razvitie turizma Karagandinskoi oblasti (Tourism development of the Karaganda region). Almaty, p 148

Tursinbaeva KS, Abdreeva ST (2013) Zonirovanie territorii po stepeni turistskoi privlekatelnosti prirodnyh landshaftov (na primere Sentralnogo Kazakhstana) (Zoning of territory according to the degree on tourist attraction of the natural landscapes (on the example of Central Kazakhstan)). In: Proceedings of the 12th interregional scientific and practical conference "Opportunities for tourism development in the Siberian region and adjacent territories", Tomsk, pp 162–166

Yerdavletov SR (1999) Geographiya turizma Kazakhstana (Geography of tourism in Kazakhstan). Almaty, p 192

Yerdavletov SR (2012) Razvitie turizma v Sentralnom Kazakhstane (Development of tourism in Central Kazakhstan). In: Materials of the V international scientific and practical conference dedicated to the 80th anniversary of university: "Tourism and recreation: innovations and GIS technologies", Astrakhan, pp 13–18

Yerdavletov SR, Artemyev AM (2007) Problemy formirovaniya turistskogo klastera v Respublike Kazakhstan (Formation problems of tourist cluster in the Republic of Kazakhstan). In: Materials of the international scientific and practical conference "Economy of the Republic of Kazakhstan in context of deepening market transformation: state and development prospects". KarSU, Karaganda, pp 277–280

Yerdavletov SR, Artemyev AM (2008) Problemy stanovleniya sovremennoi turistskoi indystrii Kazakhstana (Problems of formation on modern tourism industry in Kazakhstan). In: Materials of the international scientific and practical conference "Geographical problems of sustainable development: theory and practice". Institute of Geography, Almaty, pp 784–794

Yerdavletov SR, Alieva ZN, Mukatova DM (2014) Zonirovanie territorii Yuzhnogo Kazakhstana dlya razvitiya turizma (Zoning of the South region territory for tourism development). Bulletin of KazNU. Geographic series – Almaty, Volume 1 (38), pp 123–128

Current State and Prospects for the Ecological Tourism Development of Kazakhstan

6

6.1 Analysis of the Ecological Tourism Current State in the Republic of Kazakhstan

Ecological tourism is currently a priority for tourism and contributes to the sustainable development of the country's economy. Kazakhstan occupies the 9th place in the world in terms of territory, having unspoilt nature with many national parks and biological reserves; the country has a rich cultural and historical heritage. A research and survey conducted by the Kazakhstan Tourism Association in Germany, the United Kingdom, France, South Korea and Japan has shown that foreigners show great interest in ecological tours in Kazakhstan (Shaykenova and Mamutova 2010).

Marketing research conducted by the «IPK» (an international consulting company) showed the main socio-demographic indicators, which are presented in Table 6.1.

Based on the study, we could identify a high potential for the development of a safari trip through the steppe territories of Kazakhstan using off-road vehicles, as well as the inclusion of elements of national hunting in the tour program. The country has a variety of landscape zones for the development of this direction in tourism, consumer can meet the steppes with numerous lakes and dunes, for example in the Zhambyl region. Based on the analysis of secondary data, the following conclusions were drawn: 1. Kazakhstan has all the necessary the natural, tourist and recreational resources, while it has only the minimum necessary infrastructure to attract foreign tourists. 2. Today, a modern tourist is interested in extreme tours, where in most cases the element of extreme is the surrounding of the natural environment. In this vein, Kazakhstan has not a few mystical and secret places, with a proper level of development, they could interest foreign tourists. 3. It is possible to organize trips and attract foreign tourists to Kazakhstan with minimal infrastructure without major financial investments.

K. Iskakova et al., *Ecological Tourism in the Republic of Kazakhstan*,
https://doi.org/10.1007/978-3-030-77462-2_6

Table 6.1 Socio-demographic indicators of foreign tourists visiting Kazakhstan (official website "IPK International")

No.	Country	Share in total tourist arrivals in Kazakhstan (%)	Age (years old)	Education	Marital status (married couples in %)
1	The United Kingdom	10	35–54	Secondary and higher	26
2	Korea	8	25–54	Higher	39
3	France	6	25–54	Secondary and higher	32
4	Germany	4	35–54	Secondary and higher	24
5	Japan	1	45–65	Higher	10

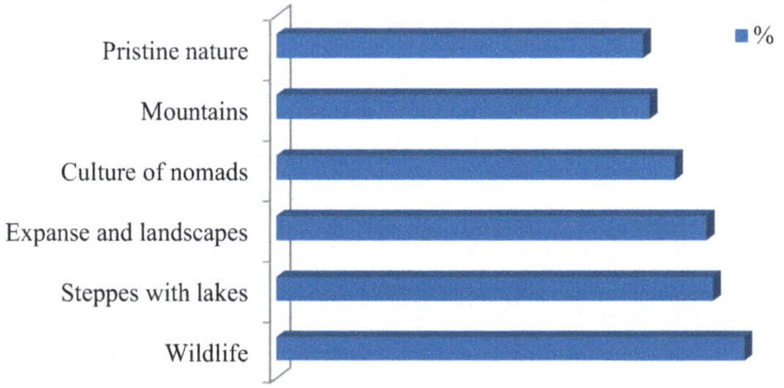

Fig. 6.1 Objects of travel in Kazakhstan preferred by foreign tourists (Report on research work 2013; Statistical yearbook of Kazakhstan 2012)

Based on the research carried out by our research group, the main consumer preferences of foreign tourists were identified, which are graphically presented in Fig. 6.1.

As a result of the conducted research, consumer preferences for types of tourism of foreign tourists in Kazakhstan are presented in Fig. 6.2.

Available tourist and recreational resources of Kazakhstan make it possible to develop ecological tourism in close connection with rural, safari, cultural, educational and equestrian-sports tourism.

Steppe safaris and historical and local history tours in Kazakhstan are the most attractive for foreign tourists. These types of tourism can be directly related with rural tourism, since, on the one hand, rural settlements of Kazakhstan are located in the steppe, on the other hand, historical and national traditions are preserved precisely in rural areas.

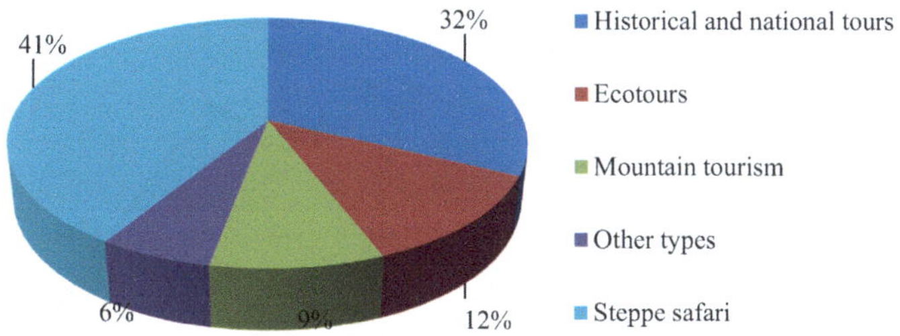

Fig. 6.2 Preferences of foreign tourists by tourism types in Kazakhstan (http://www.ipkinternational.com)

The second place in importance is given to cultural and informative (historical and educational tours) tourism. Historical and educational tours are actual in the context of interest from foreign tourists. Kazakhstan experts in the field of tourism identified 7 main areas in which the development of historical and cultural excursion programs is supposed: a section of the Great Silk Road in a territory of Kazakhstan was included there, which reveals the culture and history of nomads; more than 9 thousand historical and cultural monuments, including the mausoleum of Khoja-Ahmed Yassaui, Aisha-Bibi, Babaji-Khatun, Karakhan, Davutbek, the mysterious ancient settlement of Akyrtas, Otyrar and Sauran, underground mosque Becket Ata, Akmeshit. Historically significant places of Ordabasy, Ulytau, Turkistan, the battle of Anrakay and others. Most of the historical monuments of Kazakhstan have been restored and in the process of restoring, some have the value of the cultural heritage of mankind and are under the protection of UNESCO (Fig. 6.3).

Ecological tours took third place in the survey among foreign tourists, as their interest was attracted by the fact that Kazakhstan has unique the natural resources and protected areas. According to the latest statistics, more than 900 attractive ecological roots of a cognitive nature in the territory of national parks and botanical gardens, nature reserves have been developed in Kazakhstan. For foreign tourists there is the opportunity to watch rare and endangered species of animals, such as Tien Shan bear, barys (snow leopard), pink flamingo, kulan (onager), Przhevalsky's horse and many others (Fig. 6.4) (http://www.kaztour-association.com/).

The "Golden ring" tourist route runs through the territory of 4 national parks (Altyn-Emel, Sharyn, Kolsay lakes and Ile-Alatau).

In peculiar designated areas of specially protected the natural areas, there were equipped observation platforms—86, camp fields and tent camps—168, parking for motor vehicles—93, campgrounds, hotels, motels, camp sites—114, public catering facilities, trade—102, checkpoint—103, installed notices, signs, information boards —1148 (Final report on the development 2012).

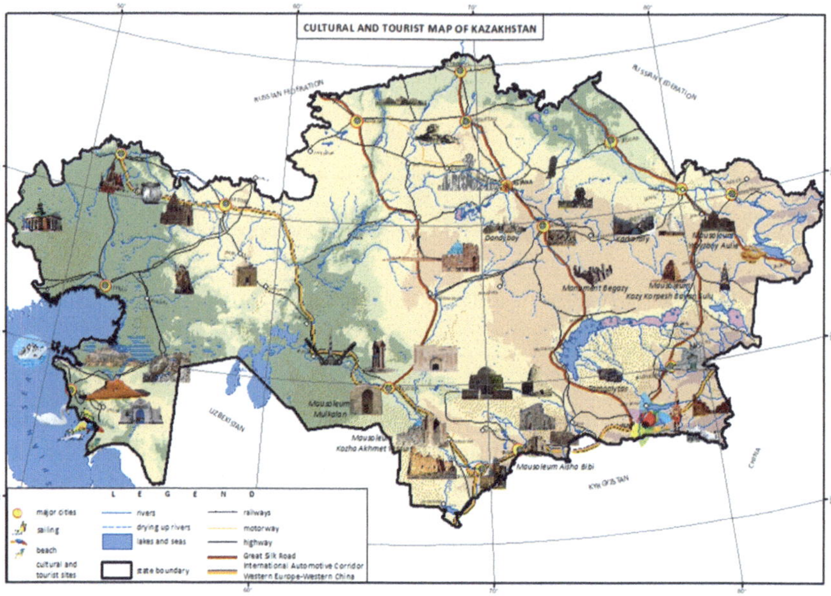

Fig. 6.3 Cultural and tourist map of Kazakhstan [compiled by the author (authors)]

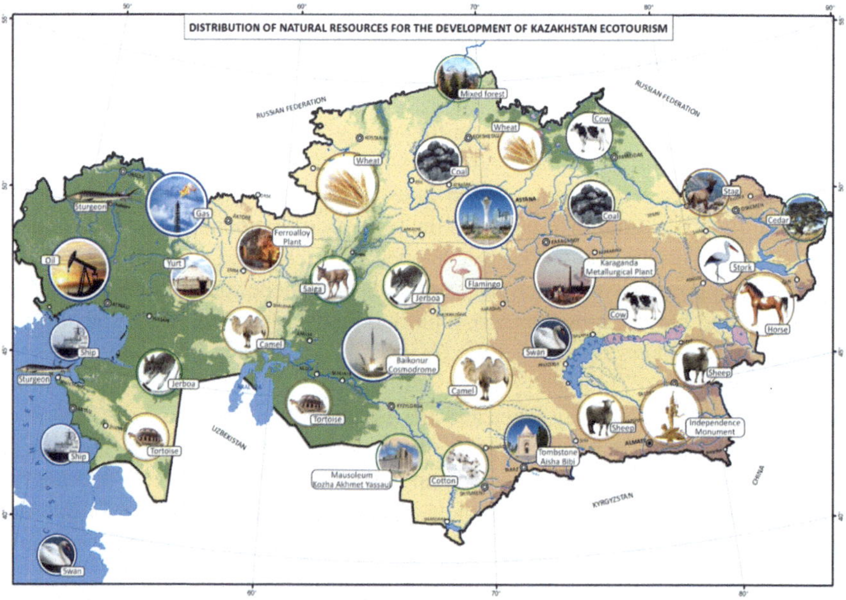

Fig. 6.4 The natural resources distribution for development of Kazakhstan ecotourism [compiled by the author (authors)]

Mountain tourism is also developing in Kazakhstan, the formation and modernization of ski resorts and recreation facilities continues. There are about 20 mountain zones in the country, commonly well-know of them is Ile-Alatau, Northern Tien Shan, Altai, Sary-Arka etc.

Thus, the main tourist product of Kazakhstan is eco-tourism, which currently has a weak competitive advantage in the international tourist market. The potential interest in this tourist product is around 9 million people (or 63% of the total potential). Ecotourism in Kazakhstan needs development and close attention (Shaykenova and Mamutova 2010).

To identify the main problems of tourism development in the country for a long period, a survey of foreign tourists was conducted. Results of the survey on foreign tourists are shown in Table 6.2.

Thus, the problem of ecological tourism development is intensification of its potential use. The problem essence is that while maintaining intact of the ecological purity on the natural areas, access to them for the broad mass population should be provided. There through, following balance of interests is ensured: environmental —protection of the natural values from the uncontrolled tourists flow; economic—receiving funds from tourists visiting the natural areas; social—education of the nature love and providing people with the opportunity of full spiritual and physical relaxation in communication with them.

The socio-economic potential of ecological tourism in Kazakhstan is practically unlimited, but its formation and development will require significant investment and costs. Creating the necessary infrastructure for ecotourism will ensure that unique corners of nature are accessible to tourists. It is necessary to create favourable conditions for attracting cash injections and investments from the private sector, in order to implement investment projects at ecological tourism facilities, for example, the construction of a tourist complex near lake Kolsai in the state national the natural park Kolsai Lakes (Fig. 6.5), the creation of a new ski resort bases in the south-east of the country and others.

The socio-economic potential of ecological tourism in Kazakhstan is practically unlimited: Shchuchinsk-Borovoe recreational zone, resort centres "Saryagash", "Arasan-Kapal", "Moyaldy", "Bayanaul", "Zhanakorgan", "Caspian", and it's not a complete list of Kazakhstan's "pearls".

Many areas of ecotourism are developing within the country (Fig. 6.6).

In order to turn Kazakhstan into a significant and well-known tourist destination by 2025, investments for 2 billion $ will be required from the state and private investments.

Today, foreigners in most cases visit our country for business purposes, which is confirmed by the data of the National Agency for statistics. Business trips make up 72% of the total trip number in 2017.

To meet the requirements of the modern market, work has begun on the formation of places of residence in the countryside, where tourists can spend the night in guesthouses or in traditional Kazakh yurts to be acquainted with local traditions and culture. Tourists visiting such places can try traditional Kazakh dishes made

Table 6.2 Main results of the survey in country (Final report on the development 2012; Report on research work 2013)

KAZAKHSTAN	
Image and positioning of Kazakhstan	For foreign tourists, Kazakhstan is presented as an original country with amazing the natural resources and many sights that need to be developed to make them more attractive to tourists However, this is a more expensive destination than neighbouring countries (Kyrgyzstan, Uzbekistan)
Based on the questionnaire, it was revealed that the main features of Kazakhstan are the natural and recreational resources, untouched pristine nature, tradition and hospitality of the nomadic people, the history and culture of the Kazakh	Key features of Kazakhstan
Main obstacles	Main barriers to the further promotion of Kazakhstan in the international market: Lack of investment in development Legal and administrative barriers (in particular, obtaining a visa) Insufficient development of tourist products and transport Lack of awareness Lack of promotional materials Many the natural resources are not accessible to uninformed tourists
Tour operators, which organizes inbound tours to Kazakhstan note that the indicators of their business activity in this market are increasing every year	The latest performance growth in the Kazakhstan market
Description and motivation of customers	Respondents noted, in general, low awareness of customers about the direction before visiting Kazakhstan The main motivations for travelling to Kazakhstan are culture, nature's addition to a business trip Today, the following categories of tourists visit Kazakhstan: couples, groups of friends or student groups, businesspersons from Russian Federation, China and South Korea
According to respondents, the following improvements should be implemented in Kazakhstan: Infrastructure, national development policy and programs, visa system, competent advertising to create the image of an interesting new country, diversity of attractions, access to the natural resources, transport system (national and international)	Necessary improvements
Competitors	Turkey, India ("sun-beach" offers) Uzbekistan (country with a very distinctive culture) Kyrgyzstan (lower prices)
Cultural tours Tradition related events Almaty—as the cultural capital Sport events, gastronomy Beach holiday	Offer of new tourist products and attractions
Further development of the hotel industry	International hotel chains, including the Caspian region (for the offers of "sun-beach" plan), while their prices must be lowered in order to compete with neighbouring countries
According to the leading tour operators of region, the main marketing efforts should be focused on expansion of international relations (forums and exhibitions) in order to improve the image of country and inform tourism industry employees about the country's capabilities	Marketing efforts (new)

Table 6.2 (continued)

Special comments	
Tourism professionals in Kazakhstan are satisfied with the performance of their work on inbound tourism In general, respondents believe that Kazakhstan has a huge potential that needs to be developed and promoted on the international market, as well as positioning Kazakhstan as an original and exclusive direction In most cases, they believe that in order to develop the tourism industry in country, it is necessary to introduce a national development policy	
INDIA	
Image and positioning of Kazakhstan	Direction is interesting for tourism, but not very well known in India, so the demand for it is low
Main features of Kazakhstan, according to respondents, are: nature, weekend tours, tourism	Key features of Kazakhstan
Main obstacles	Main barriers to the further promotion of Kazakhstan in regional markets are: Lack of awareness Lack of promotional materials Insufficiently developed tourist products
Currently, Indian tour operators do not see demand for Kazakhstan In tourism, this area is not claimed Tourist products and tour packages have not yet been developed and, therefore, country has not created the conditions for functioning of Indian operators	Latest performance growth in the Kazakhstan market
Description and motivation of customers	As a rule, Indian tourists travelling abroad with tour operators choose "all-inclusive" packages Currently, customers visiting Kazakhstan are businesspersons (booking of hotels and tickets) In general, respondents say that the awareness of customers about the direction of Kazakhstan is very limited Main motivation for visiting Kazakhstan could be culture and nature
According to the respondents, following improvements should be implemented in Kazakhstan: "all-inclusive" packages, good advertising, upscale hotels (5 *, 4 *), visa system	Necessary improvements
Competitors	Malaysia, Singapore, UAE, Thailand, Uzbekistan (Tashkent)
Tourism (inbound) Weekend tours and cultural studies Nature tours Expansion of tourist activities during business trips (in Almaty and Nur-Sultan)	Offer of new tourist products and attractions
Further development of the hotel industry	International hotel chains (4 * and 5 *) with all-inclusive packages Increase the number of international flights Development of additional business centres Development of hotel industry in the Caspian region (for the rest of "sun-beach") National transportation services

Table 6.2 (continued)

According to the main Indian tour operators, marketing efforts should be focused on: promotion in the international market; informing about the country and its main features; full training of tourism workers in direction; presence and participation in international forums	Marketing efforts (new)
Special comments	
As a rule, respondents say that the demand for Kazakhstan is practically absent except for special trips, such as business trips (although there is demand only for booking hotels and tickets) Workers in the tourism industry of India do not have enough information about Kazakhstan as a tourist destination, but they believe that the country has great potential and can become more attractive to tourists with proper information to people	
UAE	
Image and positioning of Kazakhstan	Kazakhstan is represented as a promising country for foreign tourists, but must develop its infrastructure in order to increase attractiveness and fame
Main features of Kazakhstan, according to respondents, are nature, originality, urban environment	Key features of Kazakhstan
Main obstacles	Main barriers for the further promotion of Kazakhstan in regional markets: Lack of awareness Insufficiently developed tourist products Lack of promotional materials Lack of international hotel chains
Interviewed tour operators say that Kazakhstan is not in demand by customers, so their business interests do not yet extend to the country	Latest performance growth in the Kazakhstan market
Description and motivation of customers	Tourists, travelling outside the UAE with tour operators, are usually interested in originality, authenticity, all-inclusive packages with a high level of accommodation and services. Currently, customers travelling to Kazakhstan are businessmen (hotel and ticket booking). Their knowledge about Kazakhstan before the trip is very low. Other motivations for visiting Kazakhstan can be culture, nature and weekend tours
According to the respondents, following improvements should be implemented in Kazakhstan: Attractiveness of cities "All-inclusive" packages Good advertisement Upscale hotels (4 *, 5 *) High level of education (tourism and foreign languages)	Necessary improvements
Competitors	Russia, Ukraine, Azerbaijan, Uzbekistan
Weekend tours and cultural studies Nature tours Expansion of tourist activities during business trips (in Almaty and Nur-Sultan)	Offer of new tourist products and attractions
Further development of the hotel industry	International hotel chains (4 * and 5 *) Tourist activities should be included in the tour package Business centres National transportation services

Table 6.2 (continued)

According to the main tour operators of the UAE, marketing efforts should be focused on: Promotion in the international market Relevant long-term advertising campaign Informing about the country	Marketing efforts (new)
Special comments	
As a rule, respondents say that there is no demand for Kazakhstan in the UAE today, except for special trips, such as business trips to Nur-Sultan and Almaty Workers in the tourism industry in the UAE are very interested in Kazakhstan as a tourist destination and want to get further information about the country and its main features In their opinion, Kazakhstan should expand its contacts so that as many people as possible know about the country	
EUROPE **France, Germany, United Kingdom, Netherlands**	
Image and positioning of Kazakhstan	Although Kazakhstan is not yet well known, it is perceived as an interesting country for international visitors, possessing many the natural resources and the natural beauties, however, it is closed and difficult to visit. Some tourists associate Kazakhstan with other countries of Central Asia (Uzbekistan, Kyrgyzstan)
According to respondents, the main features of Kazakhstan are culture, traditions, beauty, the natural wealth, weekend tours	Key features of Kazakhstan
Main obstacles	Main barriers to the further promotion of Kazakhstan in the international market: Legal and administrative barriers (in particular, obtaining a visa) Insufficient development of tourist products and transport Lack of awareness Lack of promotional materials Interrelation of money and quality
Tour operators, which organizes inbound tours to Kazakhstan note that the indicators of their business activity in this market do not increase and remain low. Despite this, tour operators concentrating their activities in Kazakhstan (90% of activities) speak about increasing their performance	Latest performance growth in the Kazakhstan market
Description and motivation of customers	Respondents noted that, in general, low awareness of customers about direction before visiting Kazakhstan, with some exceptions Main motivations for travelling to Kazakhstan are: Culture The natural wealth Weekend tours Places of special interest (Baikonur) Today Kazakhstan is visited by the following categories of tourists: Couples and singles Adults Groups of friends Families, where children live separately
According to respondents, the following improvements should be implemented in Kazakhstan: Infrastructure Interactions with professionals of international tourism Concentrations on national heritage Variety of events Creating an image and providing information about the country	Necessary improvements

Table 6.2 (continued)

Competitors	According to all respondents, main competing countries for Kazakhstan are the countries of Central Asia, especially Uzbekistan and Kyrgyzstan, in terms of tourism, culture, administrative procedures and low prices
Cultural tours with educated guides Tradition related events Almaty as the cultural capital Sports events, gastronomy Winter sports (skiing) Studying birds in the wild, fishing, hunting Space tourism Former nuclear zones	Offer of new tourist products and attractions
Further development of the hotel industry	Diversification of accommodation: hostels, small hotels, guest houses, as well as high-class hotels and resorts Development of direct international flights (with "Air Astana") from the main European capitals Provision of driver services who speak foreign languages
According to the leading tour operators of a region, the main marketing efforts should be focused on: International communication Information for tourism workers Organizing press tours and study tours Management of tourist activities in foreign countries Finding the right international partners for exhibitions	Marketing efforts (new)
Special comments	
Despite the fact that leading European tour operators consider Kazakhstan as an interesting tourist destination, they often do not have the opportunity and tools to promote and sell tours to European markets due to administrative barriers Leading tour operators believe that the offer is insufficient and focused exclusively on some tourist products. Prices are a bit high compared with expectations, in particular, comparing to tourist destinations in Central Asia. For this reason, the activities of tour operators in Kazakhstan are not expanding Greater enthusiasm and participation are expected from local tourism workers	
UZBEKISTAN	
Image and positioning of Kazakhstan	Kazakhstan is a very beautiful country with great potential. Thanks to the image of the former richest country in Central Asia, businessmen often visit the country (conferences, etc.) The country is considered as expensive. Tourists usually visit Kazakhstan as part of tour in the region (for example, a tour of the Silk Road). Almaty and the surrounding areas are very popular among tourists
According to the respondents, main features of Kazakhstan are authenticity, beauty and undisclosed secrets	Key features of Kazakhstan
Main obstacles	Main barriers to the further promotion of Kazakhstan in regional market are weak motivation, lack of management, low price-quality ratio, and insufficient investment in development
Tour operators, who include the visits to Kazakhstan in their proposals, note that the volume of their activity in Kazakhstan market over the past five years has been generally stable, and if it does, it will be about 1–2% per year. However, the volume of activity is not significant (not many tours to Kazakhstan are sold)	Recent changes on business efficiency in the Kazakhstan market

Table 6.2 (continued)

Description and motivation of customers	Respondents noted, in general, low awareness of customers about the direction before visiting Kazakhstan. Main motivations for travelling to Kazakhstan are business, culture, the natural wealth and travel. As a rule, businesspersons, adult tourists and tourist groups visit Kazakhstan
Respondents noted that Kazakhstan should pay attention to the following improvements: visa systems, advertising campaign to create an image of country as an attractive tourist destination, Internet marketing, lower prices for accommodation, creation of new residence places (hotels) with reasonable prices and comfort levels, as well as improving the quality of services	Necessary improvements
Competitors	Main competitors of Kazakhstan as a tourist destination, according to respondents, are Uzbekistan, Turkmenistan and Kyrgyzstan
Historical and cultural tours for travellers, the natural resources (places of special interest), weekend tours, ecotourism (nomadic life), hunting, ski holidays	Offer of new tourist products and attractions
Marketing efforts (new)	According to key operators in the regions, main marketing efforts should be focused on Internet marketing
Special comments	
Tour operators of Uzbekistan suggest that more information about Kazakhstan should be provided. They also believe that Kazakhstan has great potential, and it will become more attractive for tourists if people have status update	
TURKMENISTAN	
Kazakhstan is an interesting country; however, it requires much more advertising regarding tourist offers and places to visit Kazakhstan is considered as an attractive country for adventure and the natural resources lovers, however, the cost of movement within Kazakhstan, as well as other expenses are high, so many tourists choose neighbouring countries for recreation	Image and positioning of Kazakhstan
Key features of Kazakhstan	According to respondents, the main features of Kazakhstan are the beauty of wild nature
Main barriers to the further promotion of Kazakhstan in regional markets: undeveloped tourist products, insufficient advertising materials, and the ratio of price and quality (expensive)	Main obstacles
Description and motivation of customers	Respondents noted, in general, low awareness of customers about the direction before visiting Kazakhstan. Main motivations for travelling to Kazakhstan are commercial activities, culture, the natural resources and travelling
In the opinion of respondents, the following improvements should be implemented in Kazakhstan: advertising the country, reducing prices/expenses (prices are very high)	Necessary improvements
Competitors	Main competitors of Kazakhstan as a tourist destination, according to respondents, are Uzbekistan and other countries in the region
Disclosure of ecotourism opportunities	Offer of new tourist products and attractions
Further development of the hotel industry	According to respondents, the following types of hotels should be developed in Kazakhstan: small hotels (in the style of advertising stores with good service and traditional decoration); campgrounds (yurts—for wealthy tourists);

Table 6.2 (continued)

	guest houses/private hotels
According to the leading operators of a region, the main marketing activities should be focused on: tourism products promotion of country through participation in tourism fairs (with sociable people), advertising in the media; non-use of Soviet advertising materials, compilation of new brochures (not too much) with good pictures and in the correct English language; Creation of websites, participation in social networks (Facebook, etc.)	Marketing efforts (new)
Special comments	
According to respondents, the demand for trips to Kazakhstan is much lower than what this country deserves, for example, tourists travelling in the region have little time to visit Kazakhstan, because when choosing between Kazakhstan, Kyrgyzstan and Uzbekistan, the choice is made in favor of the latter (because of lower prices) Potential tourists in Kazakhstan are people interested in visiting the wild nature landscapes of Russia/Siberia, Mongolia, but Mongolia is an even more expensive destination, and Kazakhstan can use this as its competitive advantage	
AZERBAIJAN	
Many countries are very interested in visiting Kazakhstan due to the lack of cultural attractiveness, lack of knowledge about Kazakhstan as a tourist destination and lack of promotional materials Most of the tour operators in Azerbaijan only sell air tickets and book hotels for travelers who travel to Kazakhstan (mostly, they stay in Nur-Sultan)	Image and positioning of Kazakhstan
Key features of Kazakhstan	According to respondents, main features of Kazakhstan are the development of business relations, as well as the Silk Road
Main barriers for the further promotion of Kazakhstan in Azerbaijan market are the lack of cultural attractiveness, lack of knowledge about Kazakhstan as a tourist destination and lack of promotional materials	Main obstacles
Description and motivation of customers	Respondents, as a rule, note that the knowledge of customers before visiting Kazakhstan about this tourist destination is very limited. Main reason for visiting Kazakhstan is exclusively commercial activity. The tourists visiting Kazakhstan are mostly businessmen
According to respondents, the following improvements should be implemented in Kazakhstan: visa system, lower prices for accommodation and expansion of the international hotel chain 5 * and 4 *	Necessary improvements
Competitors	Respondents consider that the main competitors of Kazakhstan are the countries of the Central Asia region Most of the respondents plan to improve their tourism products in countries of this region that they have visited in the last 7 years
History and culture, travel, beach holidays and weekend tours are tourist products that should be developed for tourists planning to visit Kazakhstan	Offer of new tourist products and attractions
Further development of the hotel industry	According to respondents, Kazakhstan should develop international hotel chains 5 * and 4 *
Leading tour operators of Azerbaijan believe that the main marketing activities should be focused on Internet marketing and television marketing, as well as the overall presence in mass media	Marketing efforts (new)
Special comments	

Table 6.2 (continued)

Most respondents have heard positive comments about Kazakhstan Due to the lack of advertising campaigns and materials, Kazakhstan is not regarded as a tourist destination to visit Most of the respondents cooperate only with tour operators from Nur-Sultan	
CHINA	
Image and positioning of Kazakhstan	Kazakhstan is considered as a beautiful country with great potential for the development of tourist business, however, the lack of information and knowledge about it does not allow tourists to get a clear idea of this direction
The main attractive features of Kazakhstan, according to respondents, are culture, traditional values, and the natural resources	Key features of Kazakhstan
Main obstacles	Main barriers to the further promotion of Kazakhstan in regional markets: lack of knowledge, lack of developed tourist products, insufficient advertising, low awareness and insufficient investment in development
The volume of activity in Kazakhstan is low and has remained so for the past few years	Latest performance growth in the Kazakhstan market
Description and motivation of customers	Main reasons for visiting Kazakhstan are cultural values and the natural resources, while visitors are mostly singles, groups or adults
According to respondents, the following improvements should be implemented in Kazakhstan: diversification of hotels, promotion in international markets, provision of more information about tours and the country as a tourist destination	Necessary improvements
Competitors	Main competitors of Kazakhstan are other tourist destinations in the Central Asia (Turkmenistan, Uzbekistan and Kyrgyzstan)
Travels Places of special interest (e.g. casinos)	Offer of new tourist products and attractions
Further development of the hotel industry	In the opinion of tourism workers, the following hotel types should be developed: International high standard hotel chains Hostels Small hotels
According to the leading tour operators of China, the main marketing efforts should be aimed at taking part in this tourist destination in international events	Marketing efforts (new)
Special comments	
Major problem of Kazakhstan is the lack of awareness as a tourist destination in international market (action is needed to create such awareness). It is also very important to develop innovative tourist products, which include the main places of visit and places that are particularly attractive for tourists in Kazakhstan	
RUSSIA	
Image and positioning of Kazakhstan	Central Asian country with an interesting culture, unique ethnic characteristics and national colour; Mostly business travel (motivation); Main attractive features: the natural resources (for example, lake Alakol, Baikonur); Rich heritage (history, the natural resources) and attempts to attract visitors with innovative projects
According to respondents, the main features of Kazakhstan are culture, traditional values, and hospitality	Key features of Kazakhstan

Table 6.2 (continued)

Main obstacles	Main barriers to the further promotion of Kazakhstan in the international market: Lack of knowledge; Undeveloped tourist products; Lack of promotional materials; Low level of awareness; Lack of investment in development; Lack of joint promotional activities
Latest performance growth in the Kazakhstan market	Tour operators, which include Kazakhstan in their offer, note that their activity in this market is stable or has a positive trend. At the same time, only one operator claims that recently there has been a slight decrease in the number of tours sold to Kazakhstan
Description and motivation of customers	Respondents noted, in general, low awareness of customers about the direction before visiting Kazakhstan. Main motivations for travelling to Kazakhstan are culture and the natural resources, although businessmen predominate among customers Currently, Kazakhstan is visited by singles, adults, businessmen and sometimes groups of friends
Necessary improvements	According to respondents, the following improvements should be implemented in Kazakhstan: places to stay, excursion programs, promotion in major consumer markets, providing more information about tours and country as a tourist destination. Also, there should be upgraded significantly most of the standards. Kazakhstan must be positioned and developed as a brand (that is, a separate tourist destination). Standards of service should also be improved with regard to catering and entertainment/cultural institutions. In order to promote Kazakhstan as a tourist destination at the federal level, large-scale promotional activities should be carried out. More efforts should be directed to advertising of tourist products. In general, tours to Kazakhstan can be sold on a par with (for example) the golden triangle of India or joint tours in Asia
Competitors	Kazakhstan is a less interesting tourist destination compared to other Central Asian countries, so we can say that all other countries are competitors
Offer of new tourist products and attractions	Main tour operators who sell tours to Kazakhstan agree that tourists are attracted by the same tourist products and activities: Beach holiday Cultural tours Adventure travel Mountains/lakes Suggestions for other tourist products that should be developed in Kazakhstan: Acquaintance with the natural resources Business tourism Ski holiday Ecotourism Tour operators that do not include Kazakhstan in their travel offers recommend the following ways to create additional demand: Informational, educational and excursion tourism

Table 6.2 (continued)

A sufficient number of various programs for different customers (depending on social status and categories, family holidays, holidays for adults, young people, etc.) Tours around the country with a story about local national cultural features	
Further development of the hotel industry	Tourism workers believe that the following types of accommodation should be developed: 4 * International hotels 3 * International hotels Hostels Small hotels
According to the leading tour operators of a region, the main marketing efforts should be focused on advertising this tourist destination on international events	Marketing efforts (new)

Fig. 6.5 Kolsai Lakes state national the natural park (author's photo)

from ecologically clean products that meet international quality standards. Data on the use of ecotourism resources are given (Table 6.3).

The sustainable development of the tourism industry will be aimed at creating a competitive infrastructure, creating new national tourism products, and promoting them on the domestic and international markets.

The development of ecological tourism should contribute to the establishment of sustainable relationships between humans and nature. The feeling of freedom and expanse, relaxation around the nature gives us is a biologically determined need of a human as a part of living nature.

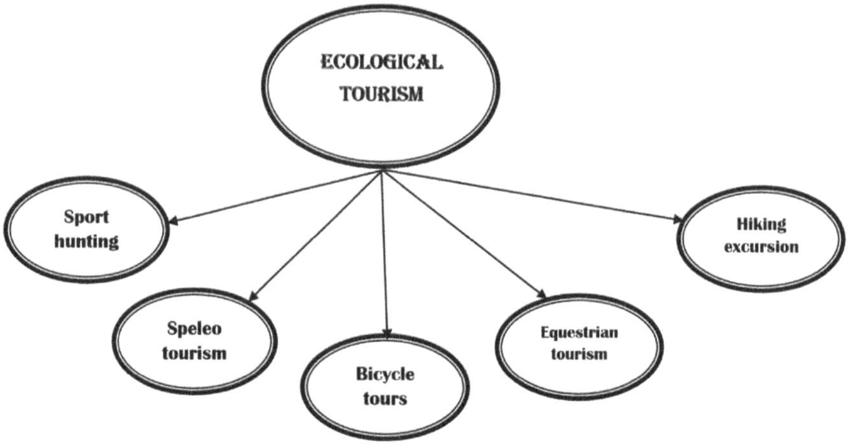

Fig. 6.6 Main types of ecotourism in the Republic of Kazakhstan [compiled by the author (authors)]

Table 6.3 Statistics of the ecotourism information resource centre (Shaykenova and Mamutova 2010)

Statistics	2005	2006	2007	2008	2009	Total
Eco websites	210	423	508	642	47	1830
Weekend tours	349	594	681	988	30	2642
Individual tours	28	65	277	360	8	738
Visiting the centre	351	457	639	492	355	2294
Total	938	1539	2105	2482	440	7504
Foreigners	632	1245	1801	2016	344	6038
Kazakhstan citizen	306	294	304	466	96	1466

6.2 Competitiveness of the Country's Ecological Tourism

According to experts of the United Nations World Tourism Organization (UNWTO), ecological tourism has become the most popular over the past ten years and is one of the instruments of sustainable development of any country, as a "direct result of world recognition and reaction to environmentally sound world practices" (Diamantis 2010).

The main purpose of ecotourism is to familiarize tourists with the natural values, environmental education and the education of people. A feature of the organization of this type of tourism is to ensure minimal impact on the environment when organizing trips, creating a network of ecological hotels, providing tourists with ecologically clean food, etc. Ecotourism includes visits to ecologically clean the natural areas (reserves, wildlife sanctuaries, botanical gardens, national parks) not altered or minimally altered by human activity. The local population with the help of tourism activities can receive not only economic benefits, but also can contribute to the conservation of the natural resources of the area, expand knowledge in the field of environmental education. In contrast to mass tourism, the organization of ecological tours can involve "local people, charitable and volunteer organizations" (Minnaert et al. 2011) which play an important role in the formation and development of ecological tourism in the rest of the world.

Today, there is an increasing demand on the world market for travels with environmental goals within specially protected the natural areas. The share of such trips for residents of America and Europe is from 20 to 60%. Meeting the increasing demand for services in this area is impossible without considering the "Environmental factor", which begins to act as an economic category (Kazakov 2009).

The tourism industry is included in the list of priority directions of economic activity of Kazakhstan. The Government of Kazakhstan pays special attention to solving a number of problems in creating a competitive tourism industry in the country. Today, ecotourism in Kazakhstan is not sufficiently developed. The scientific community of Kazakhstan in the field of tourism did not define clear criteria for understanding ecotourism; there is no consensus among experts about what it is. In these conditions, it is difficult to determine the list of planned activities for the development of ecotourism.

The rate of development of eco-tourism in the world over the past decades has been quite high and sustainable, and the high degree of integration on the tourist market contributes to its strategic development even during unfavourable economic conditions or financial crisis, when risk factors multiply increase.

According to the World Economic Forum (WEF), in 2017 Kazakhstan took the 81st place in the world tourism competitiveness index (Schwab et al. 2017) (Fig. 6.7).

Among the post-Soviet countries, Estonia has the best performance (37th place), Russia is the second, Latvia is the third (54th), Lithuania is the fourth (56th), Georgia is the fifth (70th place), on the sixth—Azerbaijan (71st place), on the seventh—Kazakhstan (81st place), on the eighth—Ukraine (88th place), on the ninth—Tajikistan (107th place), on the tenth—Kyrgyzstan (115th place).

Kazakhstan holds the strongest position in the rating on development of information and communication technologies (52nd place), on security (58th place) and favourable business climate (36th place). In terms of price competitiveness, Kazakhstan ranks eighth among 136 countries.

Based on the quality of tourist infrastructure (hotels, resorts, entertainment facilities) Kazakhstan is in 97th place. Foreign experts rated the openness of the country for international travellers as low (113th place). Among the criteria by which openness was evaluated are visa requirements. In terms of the country's visa openness, Kazakhstan ranks 114th out of 136 countries. According to the bilateral agreements on air communication, our country ranked as 121st.

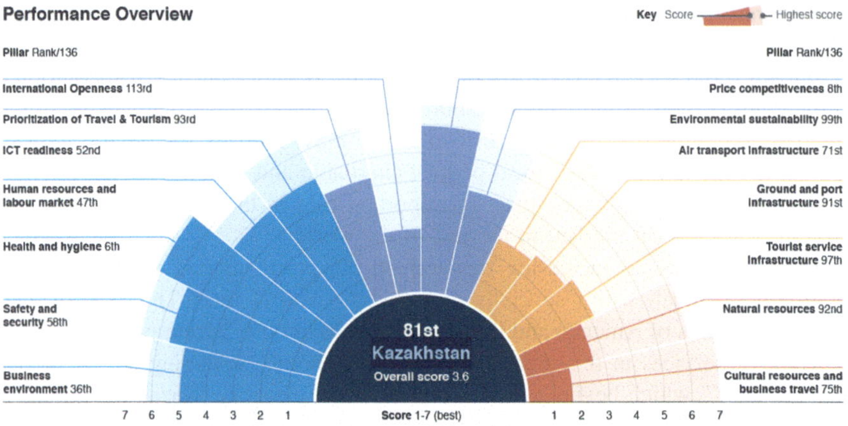

Past performance

Travel & Tourism Competitiveness Edition	2015	2017
Rank	85 / 141	81 / 136
Score	3.5	3.6

Fig. 6.7 Kazakhstan's place in the global tourism competitiveness index (Schwab et al. 2017)

The WEF experts also estimated the size of government spending on tourism development (Kazakhstan took the 46th place), effectiveness of marketing and branding in promotion the country as a tourist destination (102nd place).

According to the ecological situation (ecosystem stability), Kazakhstan took the 99th place. In the category of "Health and hygiene" Kazakhstan ranks eighth in the world.

Based on data from the World Economic Forum, Kazakhstan ranks 92nd (2.55 points) for the availability of the natural and recreational resources and 75th place (1.64 points) for availability of historical and cultural resources, while countries with highly developed ecological tourism, such as Mexico (2nd and 10th place, respectively), Costa Rica (3rd and 66th place, respectively), PRC (5th and 1st place, respectively) and many other countries are located on the top lines of this rating, which indicates the need to have rich the natural and cultural resources for development of ecotourism in country.

Ecotourism can become an important element of the Kazakhstan development, if it is considered and developed as highly profitable and at the same time environmentally safe type of tourism.

Today, there are 10 state nature reserves in Kazakhstan; 12 state national the natural parks; 3 state forest the natural wildlife-sanctuary; 50 nature wildlife refuge; 26 the natural significant-site monuments; 5 state Botanical gardens in the cities of Almaty, Zhezkazgan, Karaganda, Ridder, Bakanas village; 5 state protected areas (Aydpakelov 2017).

Today, considering environmental requirements and the natural and recreational features, in the national parks "Ile-Alatau", "Bayanaul", "Kokshetau" and "Altyn-Emel" developed master plans for the development of tourist activities that optimally determine the planned location of the main objects tourism in their territories.

In 2015, work began on drafting master plans for the development of another four specially protected the natural areas, under which improvement of the infrastructure base is envisaged, work has begun on the training of specialized personnel.

These institutions operate departments of environmental education and tourism, whose main task is the ecological tourism development, organization, conduct and regulation of cultural and educational, recreational and tourist activities in territories of parks. Eighty-four tourist and sightseeing routes were prepared on the territory of the above-mentioned national parks, the total length of which is more than 2 thousand km. In specially designated places on the territory of the parks, there are work 36 observation platforms, 58 bivouac sites and camps, 750 information stands and signs.

The following measures are being taken to further develop ecotourism in special protected the natural areas:

- processing the system for regulating tourist flow to the SPNA through selection of tour operators;
- network development on tourist routes of various directions with their subsequent arrangement and vetting;
- ensuring the security and safety of visitors and inhabitants in the SPNA;
- development of modern infrastructure for tourist activities to attract interested business entities and additional investors, tourist organizations, and local people;
- improvement of tourist routes (arrangement of recreational facilities, shelters, manufacturing and installation of small architectural forms, etc.);
- carrying out promotional activities.

In Kazakhstan, the development of economy at the expense of social service industries lags significantly behind the level of developed countries. In developed countries, up to 70% of GDP is produced in the service and financial sectors. A significant proportion of revenues provide tourism. In countries of the European Union, tourism revenues average about 5.5% of GDP (Nikonorov 2013). The budgets of countries such as Turkey, Egypt and Cyprus are more than half replenished by tourism. In the UAE, tourism is considered as an alternative to oil exports, investing huge amounts of money in its development. The same can be said about countries such as Mexico (tourism provides 33% of revenues in the structure of national budget), Norway, Venezuela, etc. These countries have the richest hydrocarbon resources, as well as Kazakhstan.

Possessing a vast territory, rich in the natural, historical, cultural and recreational potential, tourism in Kazakhstan is uncompetitive, its development occurs without a system. Over the past five years, the share of tourism revenues in Kazakhstan's

GDP did not exceed 1.9% (Turner 2018), while ecological tourism indicators are even lower in total income. Social and lawmaking obstacles lead to an inert development of ecotourism in Kazakhstan, its slow dawn the country's integration into the world tourism market. Economic and organizational reasons also hinder the development of ecotourism in our country. The list of economic barriers:

- it is necessary to have a large start-up capital to create ecological centres, and at the same time, it is necessary to consider the fact that the profitability of this activity requires a long time;
- the small amount of money injected into the creation and modernization of ecotourism infrastructure facilities by the state and the private sector does not have an excellent effect on the condition of roads and tourist accommodation;
- lack of funds to conduct an advertising campaign to attract the attention of potential tourists to visit territories that have any infrastructure for receiving, hosting and serving guests.

Among organizational barriers, there are:

- lack of the necessary legal framework for ecotourism;
- limited tourist routes in ecotourism areas and their poor infrastructure;
- a limited number of qualified specialists in the field of ecotourism who can undertake the development, organization and conduct of environmental tours;
- lack of specialized tourist organizations in the field of ecotourism;
- no advertising of ecotourism in Kazakhstan.

Ecotourism in Kazakhstan is mainly targeted at people with high-income level. There is a big role in the direction of outbound tourism, which stimulates the outflow of currency abroad. Due to the crisis, the situation in the inbound market has worsened, according to data for 2015—the number of foreign tourists has decreased by 30%. This development of events gives a chance for development of domestic ecological tourism in Kazakhstan.

In the field of ecotourism, the interests of environmental protection, the economy, culture, health, safety, international relations and intercultural communications, employment and the educational level of the population are closely interconnected (Drozdov 2010). Through the development of ecological tourism, the government solves socio-economic problems—the transition from a raw material orientation of the economy to an innovative path of sustainable development, improving the welfare and quality of life of the population, preserving the natural resources for future generations, and much more.

Thus, it can be noted that, despite the measures taken by the government to develop the tourism industry in Kazakhstan and strengthen its position in the global tourism market, domestic tourism remains an unprofitable sector of economy. Stable prevalence of outbound tourism over inbound tourism only enhances the pace of development on tourist "donation" and reduction in number of tourist arrivals. Fragmentation and inefficiency of many actions for development of tourism necessitate a more profound analysis of the industry state, identifying key problems that impede the formation of unified and positive image of the Kazakhstan tourism brand.

Certain conclusions on this issue were obtained during sociological research in the framework of research project: "Forming a country brand "Kazakhstan" on context of country's innovative development".

Purpose of study is to determine the level of attractiveness on tourism industry from the point of domestic tourist companies view and to identify the reasons that prevent creation of positive and attractive image of "Kazakhstan" for foreign and Kazakhstan tourists.

Study objectives were identification of associations connected with the country; determining the level of attitude on target audience to Kazakhstan and tourism; identifying prospects for the development of tourism industry in Kazakhstan; identification of key problems impeding the industry development.

Analysis of in-depth interviews with representatives of domestic companies showed a positive attitude towards Kazakhstan and the associations connected with it. When there was a question on what concepts Kazakhstan is associated with, the following were noted by respondents. Many of them associated the country primarily with the natural wealth and attractions: "large territory", "the natural resources", "open spaces", "steppes", "mountains", "deserts", "ponds", "lakes", "plenty of sunshine", "Caspian Sea", "Medeo" and "Astana"; with the president and people living—"Nazarbayev", "smiling people" and "Kazakh hospitality". Some also mentioned economic achievements: "a country with great economic potential", "modern country", "development".

Thus, positive associations can form a solid basis for the development of Kazakhstan travel brand.

According to respondents, Kazakhstan is an attractive place to stay. When asked what makes the country attractive for foreign and Kazakhstan tourists, the research participants noted two main points.

Firstly, Kazakhstan has a unique nature, different landscape features and interesting places that are not found anywhere in the world. According to respondents, the Medeo high-mountain skating rink, the foothills of Trans-Ile Alatau, the Asian and Kazakh way of life are attractive to foreign tourists. Foreign guests go to villages, live in yurts, learn to install them, graze cattle, drink shubat and koumiss, participate in horse races, specifically they were interested in original tourism. As a result, with the support of Kazakhstan Tourist Association (KTA), guest houses are being developed in villages of Almaty, Akmola and South Kazakhstan region.

European and American tourists are more interested in the nature of Kazakhstan, its primeval and unspoiled civilization.

Foreigners also see Kazakhstan from a business point of view—as a profitable country for investment in mining, oil and gas, and their attention is attracted by large territory.

There is a demand for mountain and extreme tourism (rafting, mountain hiking, balloons, ski tourism, helicopter tourism), especially from the Russians, Chinese and Americans.

According to the experience of tourist companies, Kazakhstan is also attractive for foreign tourists by fishing and hunting. Domestic companies offer programs to leave the island (uninhabited), where in extreme conditions; tourists can live for several days, hunt and go fishing.

Currently, at the government level there is no systematic program for tourism industry development in Kazakhstan, which would draw on the experience, professional and practical knowledge of specialists in the tourism field.

Visa and registration issues of foreign guests', arriving in Kazakhstan are topical. Mass tourism begins where it is easier to enter the country. The cost of obtaining a visa reaches about $120, while many countries establish a visa-free or facilitated regime. Examples of this are Kyrgyzstan, Turkey, Hainan island (China), which established a visa-free regime, as a result of which these areas became very popular and brought a large amount of financial flows into the country. Vietnam is developing along the same path.

Along with a visa, the entry of foreigners into the country complicates their registration with migration police. Partially this problem was solved only in Almaty, where the guest can immediately check in at the hotel for an extra service upon arrival, without losing time.

Thus, many foreign guests do not understand the Kazakhstan guest-reception system, and they often have questions about why it is necessary to spend so much time on paperwork.

Transport, air and sea infrastructure is poorly developed in Kazakhstan, the quality of roads leaves much to be desired, and access too many sights of the country is complicated. Hotels, rest houses, restaurants do not meet international standards of service.

Presence of security in country is the most important postulate in tourism. Safety is associated with food, stay, and in communication with people. In Kazakhstan, everything is very harsh, according to the respondents. People in uniform, i.e. representatives of law enforcement, border and migration services, alarm many foreign tourists. Respondents believe that tourism should become a national ideology; every citizen should understand that tourism is another type of income and desire tourists to visit their country repeatedly.

When there is a question on which regions of the country are most prepared to receive tourists, respondents noted the following:

- East Kazakhstan (lake Zaisan, Markakol reserve);
- Akmola region (resort "Borovoe");
- Nur-Sultan, Almaty as event tourism centres;
- Mangystau, South Kazakhstan region (pilgrimage and religious tourism);
- Almaty region (Ile-Alatau national park, Altyn-Emel national park, Sharyn canyon).

To the question of what they would suggest doing for transformation of Kazakhstan into the world tourist region, research participants identified the following areas:

- rendering government support in the full development of country's infrastructure: roads, electricity, communications, light aviation, lifting a ban on charter flights, launching a network of electric trains, expanding and qualitatively updating land, water and air transport;
- provision of preferences for those who acquire land on construction of tourist complexes, based on the experience of Turkey (for example, the construction of hotels was carried out at the expense of public investment and was later given to business management);
- an increase in investment of higher education system by profile and preparation on professional guides;
- to minimize all administrative barriers, reduce the volume of state control and move to "partner" level, facilitate the visa regime and registration of foreign guests;
- to establish quality work of regional tourism departments with tourist companies' representatives;
- to establish a communication system with outside world: to develop an advertising campaign about Kazakhstan and broadcast a series of commercials in markets of Europe, the USA and other countries;
- to open tourist departments in Kazakhstan embassies that provide background information about country, since not all foreigners trust the Internet sources and want to make sure of the accuracy; so, there is a need to actively promote city and other official websites about Kazakhstan in various search engines;
- to establish a system of training seminars for tourist company employees wishing to improve their qualifications on a monthly and free basis.

Thus, the study results confirm statistics on the low level of Kazakhstan tourism development. The country remains as a tourist "donor" for countries that are actively attracting foreign tourists.

6.3 Perspective Directions of Ecological Tourism Development in Conservation Districts

Ecotourism makes a major contribution to nature conservation through the support of specially protected the natural areas, which play sufficient role in maintaining the ecological balance of the planet. However, the natural areas are at risk of pollution not only because of industrial development, but they are also threatened by an extraordinary influx of visitors, because of intensive tourism development in general.

If earlier the main task of protected areas was to protect sites and territories where there were only unique landscapes, rare and endangered species of plants and animals, the realities of the modern world set a much broader task—to preserve and, if its possible, restore and develop the aesthetic, recreational, cultural and historical qualities of the natural landscapes which are the basis for development of ecological tourism. There was a need to reveal and show people all the uniqueness and diversity of the primordial nature that has been preserved in specially protected areas, and on its example to cultivate a careful attitude to it.

One of the best examples on ecotourism development as the most important sector of economy is Kenya. Most foreign visitors come to Kenya for ecotourism purposes. The government of this state pursues an active policy in the development of ecotourism. Success in ecotourism development is mainly based on the presence of well-established system for protected areas. Almost a third of Kenya's territory is allocated to national parks and reserves. Since 1977, there is a law on the complete prohibition of hunting. In Kenya are the highest peaks of Africa, the so-called mountains of Kenya and Kilimanjaro. The national park "Lake-Nakuru" has been created for protection of several million pink and ordinary flamingos. The most popular reserves in Kenya are Massan-Mara. For tourists, they are allowed to observe animals, photographing and video filming (Samoilenko 2006).

Previously, most of the funds received from ecotourism went to the state budget, bypassing protected areas and local community. Currently, the Wildlife Service has been established in Kenya, authorized to manage the SPNA, and to rationally distribute all funds. As a result, the bulk of revenues began to remain in the field, which greatly alleviated financial problems of protected areas. Representatives of the travel industry in Kenya designed a special course on ecotourism organization.

Thus, ecotourism development occurs using the natural capital. To date, the natural environment has not undergone significant anthropogenic changes. This means that in this country, ecotourism can develop in full measure.

Ecotourism has received significant development in the Republic of South Africa. In 1994, the republic joined the World Tourism Organization. This year was a year of drastic measures in the field of tourism. Tourist Council of South Africa provides special training for tourist guides due to the great diversity of country's the natural conditions: from the desert to tropical rain forests.

Currently, ecotourism has become the most profitable and rapidly developing type of tourism in South Africa. UNESCO protects many of the national parks of the country. Types of ecotourism in country are diverse—it is hiking, botanical excursions, aquacades and crocodile nurseries are popular. Nature of the country did not experience large anthropogenic pressures; therefore, significant areas of the natural landscapes that are suitable for organizing ecological tourism have been preserved here (Samoilenko 2006).

In the North American region, ecotourism is represented by a fully developed SPNA system. Ecological tourism is excellently developed in Canada. It is no accident that in 2002, which was declared by UNESCO as the Year of Ecotourism, the World Ecotourism Summit was held in Quebec, in which more than 1000 delegates from 132 countries took part. The organizational centres of nature tourism are primarily national parks, the most well-known of which are Banff, Dinosaur, Glacier, Ioho, and Buffalo.

Number of tourists who are willing to visit these places constantly growing; therefore, the recreational infrastructure is being improved: hotels, mountain huts, ski slopes are being built, and cycle-pedestrian trail system is being developed.

The United States, which sends the largest number of ecotourists abroad in the world, also has a huge domestic ecotourism market. More than 10% of the US area is covered by protected the natural areas. The most visited national parks that contain unique the natural features-such as the Grand Canyon in Arizona, or are located near large urban areas (Yosemite, Shenandoah, and Great Smoky mountains).

With the creation of the world's first Yellowstone National Park in 1872, a wildlife protection program in the United States began to operate. In addition to a particularly complex network of protected areas, a system of non-state reserves has been developed, including more than 1 thousand reserves.

Ecotourism is a very profitable business in Alaska. Alaska is home to the largest U.S. the natural reserve, the Arctic national faunal reserve.

The Northern forests of the United States border the Great lakes, which are also ecotourism sites. The Hawaiian Islands also have unique protected areas. In Hawaii, the highest active volcanoes in the world are located, unique species of water and coastal birds live here, exotic plants, 90% of which are not found in other parts of the world.

Ecotourism is not only a means of nature conservation, but also contributes to the improvement of settlements infrastructure located near passage of ecological roots.

Central and South America is rich in a unique the natural environment, many animals and plants live and grow here. On the islands of the Caribbean, there are more than 100 marine protected reserves.

On the island of Puerto Rico, foreign tourists have the opportunity to visit the Caribbean National Forest, where they can get acquainted and observe the large number of wild plants, which make up 70% of the total plants of the island. For example, on the island of Dominica, tourists can enjoy exploring the greatest collections of rare orchids in the world. For the convenience of tourists, the government of the island state developed walking routes around hot springs and waterfalls provided the entire necessary infrastructure for these territories. Local tour operators have developed an underwater route to the sacred volcano Soufriere to increase the flow of foreign tourists.

Capital of Caribbean ecotourism is the national parks in the Belize state. National parks are protected parts of the Barrier reef—the longest on Earth.

Germany is one of the leaders in the development of ecological tourism from the countries of Central Europe; in addition to high standards of quality of service, there are ancient forests, unique the natural landscapes that attract a significant number of tourists. In France, rural tourism is developing well, one of the areas of ecotourism, which aims to familiarize tourists with the actions necessary to preserve the natural balance and protect the environment, as well as familiarity with the culture and tradition of French cities. Great Britain boasts one of the largest nature reserves—the Scottish mountains, in addition to it, there are several thousand more nature reserves.

Ecotourism in the UK is an additional means of economic development on local population and land restoration.

Finland rightfully occupies one of the first places among European countries with the most favourable environmental conditions. This is facilitated by state policy, in which respect for the natural resources and minimal human influence on the ecological environment are elevated to the national idea rank. Ecological tourism is a kind of visiting card of a country.

In Finland, there are more than thirty national parks with a total area of 8170 km^2. Some of them are still not fully understood by man until now. Promotion on the parks' territory of any vehicle type is prohibited, only walking routes are allowed. However, among the tourist programs of the country there are those that provide flights to hard-to-reach places on aeroplanes or helicopters.

The Asian continent is fascinating for ecotourism. They are stand out particularly in South Asia, India and Nepal, in East Asia—China, Malaysia and Japan, in South-East Asia—Thailand, Indonesia, and Vietnam.

Ecotourism is very well-developed in Australia; the main stream of tourists is adherents of sustainable tourism. The unique flora and fauna, the presence of endemic species is of great interest to visitors. On the continent, there are more than 3 thousand national parks and reserves, which have a highly developed infrastructure, specially equipped tourist sites and routes, many tourist routes for different age groups. One of the most well-known national parks is Kakaduin the north of Australia. Local aborigines manage the Park. Routes are lasting from 2 to 5 days.

On the island of Tasmania, about 30% of the territories are on the World Heritage List. Tourists are attracted here by absolutely wild mountainous terrain.

An information resource centre for ecotourism has been established in Kazakhstan, as well as a Committee for forestry and wildlife, in some areas of local authorities managing SPNA. All these organizations focus their efforts on solving the problems of developing ecological tourism. However, despite this, in many sites of SPNA in Kazakhstan, tourism often develops beyond any dependence on nature conservation. Private travel companies of Kazakhstan are mainly engaged in the development of elite ecotourism for wealthy people from abroad, which brings them a huge income.

Recently, the local authorities of Kazakhstan are more and more actively concerned with problems of sustainable development in territories, where they are facing the problems of ecological tourism associated with sustainable development in tourism industry. Concept of sustainable development in the Republic of Kazakhstan, strategic environmental plans developed the Ministry of Investment and Development of the Republic of Kazakhstan, the Ministry of Agriculture of the Republic of Kazakhstan, and through indicating the development of nature-saving tourism in future, such as ecological tourism.

Experience of foreign the natural national parks shows that the entrance fee to park zone is usually very low, which allows them to be open to almost all categories of people. In our country, due to financial problems, it is planned to introduce differentiated admission fees for foreign and domestic tourists, as well as, depending on the nature of services provided to tourists in the park during their visit. For example, the entrance ticket for visiting Altyn-Emel national park in Kazakhstan was about $3 for 2017, and for foreign tourists the cost was about $20.

A differentiated approach in the cost for local and foreign tourists is very common in international practice. For example, the Galápagos national park charges foreign tourists a higher fee than the citizens of Ecuador, and does not charge it at all from residents.

As a rule, the development of ecotourism sets itself the following tasks:

1. Environmental education, improving the culture of relationships with nature, developing environmental standards of behaviour in the natural environment;
2. Restoring the spiritual and physical strength of a person, providing a full rest;
3. Nature protection, minimization of damage to the natural environment;
4. Promoting the socio-economic development of individual regions and States as a whole.

These objectives form the basis for functioning of any SPNA around the world. Kazakhstan's SPNA are not fully implemented, and sometimes they do not always fulfil them. In addition, there is no method of organizing ecotourism, namely: the spontaneously used eco-routes do not have the appropriate equipment and information support; the system of training for reception and environmental education of visitors has not been established.

Organization of excursions and environmental education are of great importance in the field of environmental conservation in national parks. Tourists visiting the national park should be acquainted with its functions, with what it protects, why it exists, what boundaries it has, as well as with the rules of behaviour on its territory. Such a well-developed information program containing the rules for visiting the natural park is a key element in ecological tourism development within the park area.

Today, experts in the field of tourism note that the tourism industry as a whole has a certain negative impact on the environment. Nevertheless, in the case of ecotourism, this effect is minimized due to the presence on the natural park territory of specially equipped routes, which are called "nature trails", or "ecological paths". Considering the interests of visitors, such trails may have different purposes, length, etc.

For the first time the concept of the natural or ecological trail appeared in the United States. In the early twentieth century, forester Benton McKay proposed the establishment of something like a "reserve for pedestrians"—to lay a path along the Appalachian range. By 1922, the footpaths through all Appalachians from Maine State in the north-west and to Georgia in the southeast was ready, its length was 3300 km. Moreover, immediately it became a favourite place for recreation and communion with the nature of many Americans in the eastern part of the United States (Suvorov 2017).

Any use of protected areas as objects of ecotourism bring a certain income from tourists visiting this area, part of the income from which, with proper planning and development, should be directed to:

1. Marketing and promotional activities to attract more visitors;
2. Environmental protection, stimulating the development of related infrastructure, conducting research work;
3. Educational activities and training of professional guides (including those from the local population);
4. Development of centres of environmental education, development of communication and logistics, and route trail equipment.

The processes of globalization and the integration of Kazakhstan into the world community, gives the country a unique opportunity to use the unique nature, historical and cultural objects and recreational resources that are not yet covered by the processes of urbanization and intensive agricultural production for the development of ecotourism (Fig. 6.8).

Today, as part of the state order for national parks, master plans have been developed, which also allow optimal determination of planned locations for main tourism facilities on their territory, considering environmental requirements and the natural features.

To implement the current master plans for ecological tourism development, 302 land plots with total area of more than 11 thousand ha were provided for capital construction of tourism facilities on long-term use and 161 land plots with total area of 422 ha for construction of seasonal tourism facilities on the SPNA.

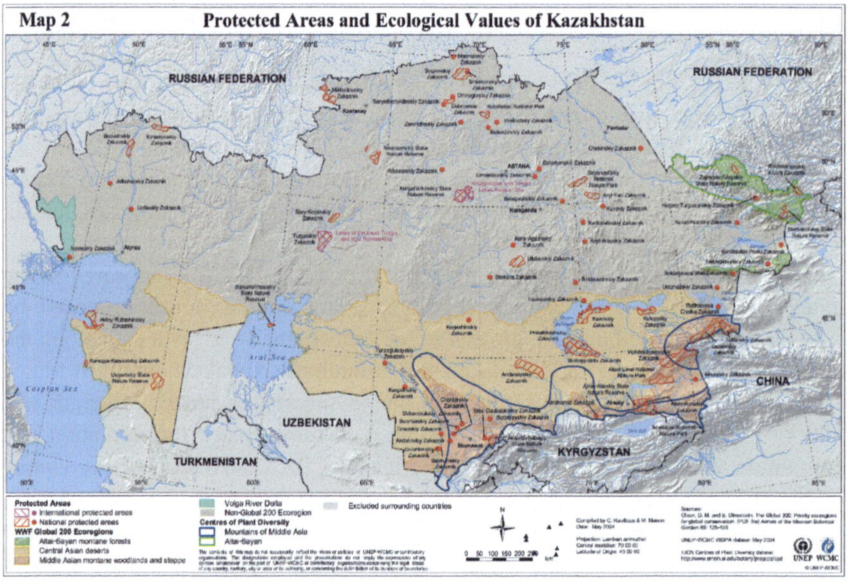

Fig. 6.8 Placement of specially protected the natural areas of the Republic of Kazakhstan (www.unep-wcmc.org, 2004)

Present day, there are 220 tourist routes and paths in specially protected the natural areas, in which reconstruction and settlement works are carried out. The number of guides, who organize excursions for SPNA visitors, is 46 people across the country. In 2016, there were 4997 excursions conducted in the SPNA (Aydpakelov 2017).

In 2017, during the international exhibition EXPO 2017, the number of foreign visitors to the SNNP of Kazakhstan was 7064 people, the amount of budget revenues was 1.7 million tenge, while the amount of income to special account of national parks was 4 million tenge.

The number of visitors to specially protected the natural areas is constantly growing (Fig. 6.9).

Today, ecological tourism development in Kazakhstan is an important source of income from the conservation of nature and environment in SPNA. Following sources of income from ecological tourism development can be distinguished:

1. Payment for environmental management in the form of admission, payment for the use of environmentally friendly transport and accommodation, hunting trophies in the form of photographs, handicrafts and souvenirs of local population and others.
2. The rent of individuals for provision of services in form of food, housing, transport, and guide.
3. Percentage from the sale of products to tourists visiting this territory.
4. Taxation of goods and services.
5. «Multiplicative effect».

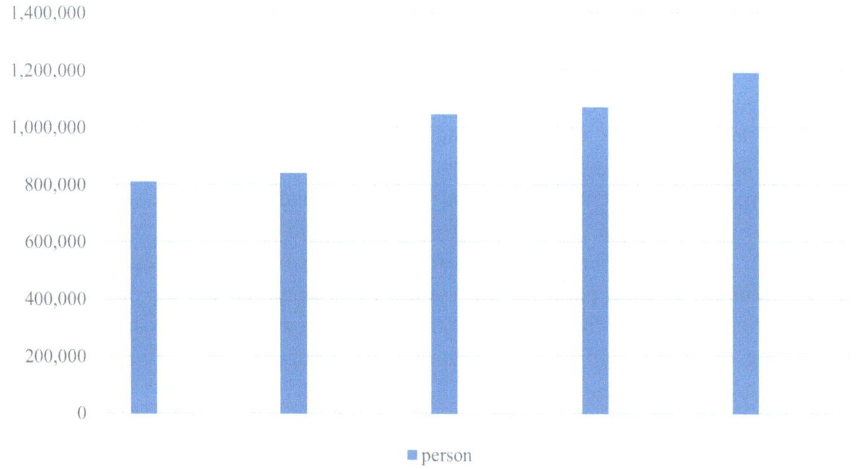

Fig. 6.9 Total number of visitors to the protected areas of the Republic of Kazakhstan [compiled by the author (authors) based on the materials Aydpakelov 2017]

The development of ecotourism is aimed at preserving the biodiversity of nature, and at the same time bring some profit from recreational activities. It depends on a number of conditions:

1. For the development and implementation of sustainable tourism in protected areas, the management of protected areas must have legal authority.
2. Protected the natural areas visited by tourists should be characterized by the presence of interesting wildlife components that can attract visitors.
3. The pricing policy should consider the solvency of consumers, the collection of payments depends on the cost of using the environmental facility.

Experience shows that when the local population is involved in ecotourism projects, such violations as, for example, indiscriminate cutting down of trees or poaching are significantly reduced.

It is important that the local population itself take an active part in ecotourism activities.

In Kazakhstan, the negative consequences of ecotourism development are not so problematic, but rather we can talk about the consequences of mass tourism in some vulnerable habitats of wild animals. However, they are sufficiently tangible. For example, amateur ornithologists to photograph the rare red-book birds of Kazakhstan on their nests are closing in to them. It should be noted that at present, most species of animals are in dysfunctional state due to the violation of their habitat by humans. All this makes today think about the ways on rational development of ecological tourism from the future perspective of Kazakhstan.

In the strategic plan of Kazakhstan development until 2025, the government of the country sets a task to increase the area of SPNA to preserve endemic, rare and endangered species, unique reference sites and the natural ecosystems. In the period until 2025, the government of the country plans to create two national parks, three the natural reserves and bring the area of specially protected the natural areas of Kazakhstan to 10–12% among the country's area. Kazakhstan promotes the policy of preserving biological diversity through increasing forestland area and bringing it to 4.8%, preserving the objects of state nature reserve fund, conserving and increasing the number of wild animals, including rare and endangered, development of game breeding in unwitting and semi-volunteer conditions, preservation of fish species and betterment of commercial fish farming and aquaculture.

To date, the government of Kazakhstan, as part of the state program for the development of national parks of the country, has prepared master plans for the optimal determination of the location of the main tourist sites in the national parks, considering environmental requirements and the natural features. Every year the number of visitors to national parks of Kazakhstan increases.

6.4 Development of Ecological Tourism in Rural Areas

As part of studying the issue on rural development, it is necessary to investigate the content of "rural territory" notion. Among a number of scientific literature, rural areas are understood as territories for which the characteristic feature is a small population, where a large proportion of manual labour prevails, and a significant part of the population is engaged in agriculture.

According to Russian researchers, "the village is a Slavic word, which meant the settlement where the landowner was located, and the surrounding places were called villages. At the beginning of the XXth century, the village was considered as a settlement, where there was a church. As a rule, several hundred people live in the village, mainly engaged in agricultural work. Here cultural traditions and labour skills are passed from father to son, from mother to daughter. All residents know their neighbours; there is a high level of conformism and social control" (Blinova et al. 1999).

Rural areas are territorial units (districts, settlements) within which the area community is formed, and characterized by specific rural lifestyle, land and other the natural resources become origin and factor of the social and economic activities concerning people. From the traditional point of view, rural areas tend to develop agricultural industries, and the final product of which serves as the material basis for territory development and source of its income.

One of the researchers in this field, the Russian author Zaytseva remark that "peculiarity of rural territory is determined by the fact that it is a source of many benefits: food and agricultural raw materials, pantry of the natural resources, place of residence, recreational facility, place for waste assimilation" (Zaytseva 2013).

From all the above, we can conclude that rural areas are a multi-functional unit with sufficient resource potential, the realization of which will ensure its territorial development.

Based on semantic and contextual analysis of information related to the concept of "rural territory", the following conclusions can be drawn:

(1) two unified concepts are widely distributed—"rural territory" and "rural area", but the second, although important for ecotourism, but hardly applicable for research and management purposes because it has a general character and implies the entire geographical space located outside the borders of large settlements—cities;
(2) according to the most interpretations, typical rural areas have not a high population density and are characterized by small populations of settlement, but considering the demographic approach, as exception can be foothill areas of south-eastern and southern Kazakhstan, characterized by high population density;
(3) in accordance with the model of main concepts correlation, three main groups of ecotourism varieties should be distinguished by territorial affinity: (a) dominant—ecotourism as a combination of tourism types in rural areas and water areas (A), including rural tourism (A1) and agritourism (A2);

(b) subdominate—ecotourism as a combination of tourism types in rural areas (B), including in specially protected the natural areas; (c) secondary—ecotourism as a combination of tourism types in urban areas and water areas (C);

(4) there are four main approaches to the allocation and management of rural areas in Kazakhstan (land use, demographic, agricultural), according to which their composition should be considered: (a) as a system of administrative-territorial units "aul (village)", "rural locality", "aul (rural) district", "rural area" (administrative approach); (b) as a complex of 9 categories on the land fund in Kazakhstan, where the land of specially protected the natural areas, health-related, recreational, historical and cultural purposes are especially important for ecotourism; lands of forest and water resources (land use approach); (c) as an aggregate of rural population in Kazakhstan, area prospects where they live for ecotourism development is determined by provision of domestic consumers with ecotourism product, unoccupied labour resources, and comfort in terms of the natural, ecological and socio-economic properties for population and guests (demographic approach); (d) as a system of agricultural land use in Kazakhstan with different level of development on agricultural production and rural entrepreneurship, the predominance of crop production in Akmola, Zhambyl, Kostanay, North Kazakhstan and South Kazakhstan region, livestock in Aktobe, Atyrau and Mangystau regions or about the same ratio in Almaty, East Kazakhstan, West Kazakhstan, Karaganda, Kyzylorda, Pavlodar regions (agricultural legal approach).

To meet the challenges of sustainable rural development, world experience demonstrates the tourism development, in particular ecological, as one of the options for preserving and maintaining rural regions.

It is necessary to develop new directions in tourism at the level of regions, rural areas; it is worth paying attention to the development on such types of tourism that would allow using the socio-economic, the natural resources, ecological potential of the country with a dynamic approach to sustainable development.

Main global trends in the tourist services' development can be identified as areas that are associated with ecotourism. Currently, Austria, Slovakia, Germany, the Czech Republic, Poland are intensively increasing their capacity to receive ecotourists. Thus, in Austria, up to 10% of peasant estates provide accommodation services. A number of European states, under the auspices of the European Council, united in the Central-Eastern European federation for development of rural green tourism. The increased interest of tourists to the countryside is associated with a change in conditions of modern people existence (Agritourism: experience, problems, solutions 2012). In international practice, the leaders in the field of rural tourism are Italy, Spain, France, Bulgaria, and Belarus.

Tourism in rural areas, being a specific territory with "multiplicative effect", stimulates the development of such interconnected sectors of economy as agriculture, trade, transport, communications, production of consumer goods, as well as the social sphere. At present, in Kazakhstan, an urgent task for tourism is the development and implementation of new socio-economic models, which should be based on the principle of economic efficiency, combined with implementation of social problems inherent in tourism. At the same time, tourist services should be equally available to all segments of population: its economically active part—on market terms, and to socially unprotected categories of citizens through the new economic mechanism of social tourism. Only mass and socially oriented tourism can be the concept basis on physical and spiritual recovery of the nation.

Ecological tourism in rural areas involves: (a) tourists temporarily stay in rural areas for the purpose of recreation and/or participation in agricultural work, with direct participation in the protection of the natural and historical and cultural environment; (b) ecotourism direction, to create a comprehensive tourism product, the natural, cultural, historical and other resources of the countryside are used.

A prerequisite for its formation is the availability of tourist accommodation facilities, specialized or individual, which must be located in rural areas without industrial and multi-store buildings (Fig. 6.10).

In terms of price-quality, Kazakhstan tourism products generated in rural areas cannot yet compete not only in the international market, but also with the neighbouring countries of the Eurasian economic union. The main orientation of Kazakhstan's tourism sector is aimed at the domestic market, which, however, does not exclude the development of certain high-quality tourist products specifically in the foreign market. International experience shows that in many countries of Europe, where tourism in rural areas has received quite a large development, it is mainly the domestic tourism sector, although the share of foreign tourists is gradually increasing, for example, in Italy, France, and Germany.

The following types of tourism are usually practised in rural areas:

Recreational leisure activity combining active and passive forms: tourists will be able to enjoy outdoor activities such as horseback riding, help in harvesting farms, walks in the surroundings by land and water transport; as well as passive species

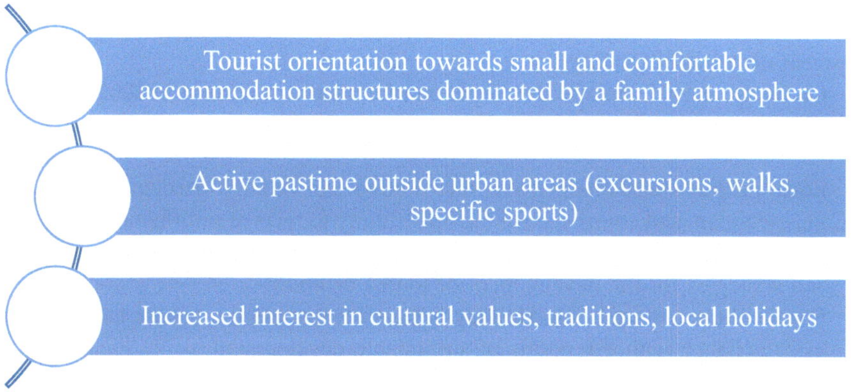

Fig. 6.10 Tendencies, which are favourable for rural tourism in the formation of tourist flows [compiled by the author (authors) based on the materials Centre for the development of tourism in the Sverdlovsk region 2011]

such as fishing in nearby rivers and lakes, acquaintance and participation in cultural events, acquaintance with history and tradition, with the folklore of the local population, forest therapy and much more.

Gastronomic tourism in the countryside is a type of holiday during which the tourist is acquainted with the farming sector of the region, it is supposed to participate in harvesting and eating berries, mushrooms, vegetables and fruits in the ecologically clean regions of the country. Moreover, this acquaintance with the peculiarities of local cuisine, culinary traditions, to try a dish or product that is unique to a visitor. As a souvenir, tourists most often take ready-made recipes for dishes that they had a chance to try.

Educational tourism is a trend in tourism activities aimed at providing the target audience with various interactive forms and types of services combining recreation and educational activities. Educational tourism in the rural area is aimed at training tourists with the types of work that are necessary for living in a "classic" village. For example, this may include training in clay and pottery craftsmanship, folk painting, woodcarving, weaving vines, and painting ceramics. The educational tourism program may include training in folk cooking, planting, etc.

Agricultural ecological tourism or green tourism is a trip to a village located in a clean and untouched area. Often without a telephone, television, the Internet and other forms of communication and with well-designed excursion and entertainment programs, a person can truly appreciate all the charms of life without the use of modern electronic devices.

The main consumer (target group) of rural tourism in Kazakhstan are: companies of young people (most often students); people of pre-retirement age; people who love sports and engage in amateur level; researchers whose field of study is related to the countryside.

Some Kazakhstan researchers in the field of tourism believe that going to the countryside does not make much sense, since the majority of urban population, such cities as Almaty and Nur-Sultan, has country houses. However, the majority of authors agrees that summer residences will create a certain competition for development of "rural hospitality".

Currently, the principle of development of ecotourism regions in Kazakhstan is applied through:

- Local people. Residents increase their knowledge through participation in various seminars and courses, have the opportunity to receive information about the experience of foreign countries, get the opportunity to exchange ideas on policies and strategies for the development of tourist centres;
- Local culture and tradition. Holiday houses for tourists use "green technology" in compliance with the norms of sustainable development, using local culture and history in the design, considering foreign experience to increase the comfort of guests. Traditional dishes are prepared, such as Beshparmak, Kuyrdak, Nauryz-Kozhe and other local products. It is also offered to watch and take part in various national sports: kokpar, tenge atu, kiz kuu, baige and others;

– Benefits for the environment. Rural communities provide tours and operate within the framework of the environmental law. In addition, communities leave a portion of their income for community development and conservation.

Strategies for involving the local population on ecological tourism development in regions are very diverse:

Option 1. Due to the increasing interest of Kazakhstan regions population on excursion and health tours in rural areas, the offer accommodation for tourists lags demand. In the Kazakhstan countryside, in the villages there is a huge shortage of comfortable hotels, even 1–2 stars, and in most cases, the role of tourist accommodation is played by guest houses of the local population that do not meet the requirements and quality standards.

Large investments are required for the construction of new tourist accommodation facilities, and in most cases, Kazakhstan entrepreneurs do not want to inject financing into enterprises with long-term payback prospects. Accessibility of price will allow more citizens to relax in the nature lap, and villagers—to receive additional income.

Rural residents have the opportunity to provide a wide range of services for tourists. Each participant in the tourism process in rural areas has the opportunity to receive additional income in addition to the main activity related to the provision of services and goods in the field of tourism. The range of services and goods may include such components as meals and accommodation in a guest house, rental vehicles (car, horse, swimming equipment and others), sightseeing activities, leisure and recreation, additional services.

The main motives for choosing a holiday in the countryside are a nostalgic rural atmosphere (many as a child came to visit grandparents in the village, which is the reason for nostalgia), participation in agricultural work and gardening, collecting gifts of nature (photo collages, herbarium, souvenirs), care and games with animals. At the same time, for urban residents, the attractiveness of this type of holiday is to reduce cash and time costs (lack of borders and proximity of holiday spots, when it comes to holiday inside the country), fresh air, solitude with nature, ecological clean and cheap food, calm recovery of physical and spiritual energy.

For Kazakhstan, ecological tourism in the countryside is a relatively new area, as the systematic development of this area has only recently begun. The main factors contributing to the development of rural tourism in Kazakhstan are favourable climatic conditions, the richness and variety of the natural resources is not too tied to seasonality, favourable environmental conditions, an attractive agricultural landscape, the presence of small and medium farms that can host tourists with relatively small investments, an interesting history and the rich culture of the nomadic people, necessary infrastructure.

The following areas can be identified as the main reasons for choosing rural tourism in Kazakhstan:

1. The need for improvement in various climatic conditions recommended by the doctor;

2. The established way of relaxing in the countryside for a certain category of people, regardless of the availability of funds, for example, due to family or other traditions;
3. Proximities to nature and the ability to spend more time in the fresh air;
4. Participations in local holidays and entertainment, such as Nauryz, Kurban Bairam and others, communication with people of a different lifestyle, the opportunity to learn the cultural characteristics of the people;
5. The real opportunity to join agricultural work for your pleasure.

One of the main conditions for the successful development of tourism is the state of the environment. Tourism cannot exist without a natural basis—the natural environment and makes special demands on the state of the landscape. Nature is the main resource of tourism. Today, the success of the popularity of most tourist centres depends on the cleanliness of the environment, effective measures to protect it and the identity of the local culture.

There are two main types of rural accommodation for tourists:

- Rent a small house or cottage, rooms in a guest house located in picturesque areas, surrounded by unique nature;
- Accommodation of tourists on a farm, in a rural house directly in the family of the owner of the house. This allows guests to "feel" the style of village life, make contacts with new people, learn interesting customs and the history of the pedigree of the owner of the house, take part in household chores and eat at the same table. In many rural houses of Kazakhstan, one or more rooms are especially reserved for guests, they are properly cleaned, and there is bed linen and necessary things. And this is connected not only with tourism, first, it is a tribute of Kazakh to their tradition, the Kazakh people are well-known for their hospitality, and there is always a place in the house of the Kazakh for an overnight stay for a traveller or unexpected guests.

The ways of placing one type or another put forward certain requirements in front of the owners. Therefore, rest in the hayloft involves the creation of an appropriate barn interior. To ensure the safety of guests from snakes, rodents and other insects, the owners of the guest houses arrange the attics in such a way that it meets high comfort with elements of authenticity.

Option 2. The involvement of local population in ecotourism development is one of the basic principles of its development. Favourable conditions must be created for local population to be interested in protecting the natural resources, receiving tangible social and economic benefits.

Attempts to develop tourism without the participation of the local population often lead to negative socio-economic and environmental consequences. The local community should play a big role in forming a positive response in the development of tourism, without solving the problems of a socio-economic nature in rural areas; it is hardly possible to stop the destructive ways of using the natural resources. The participation of the local population can guarantee that a significant part of the income from the development of ecotourism will remain in this area.

The need for relationship of protected area management and their economic activities with activities of local population is accentuated in the World Conservation Strategy (JUCN 1980). The International congress on national parks (1982) emphasized the need to involve local population in the planning and management of protected areas, calling for increased support for population living approximately protected areas through such measures as decision-making and the strategy for the further development of the area, their participation in the distribution of income, and the improvement of the education level of residents.

Benefits that can be obtained from the development of tourism in the countryside:

- To increase the general level of rural welfare, it is necessary to pay close attention to the development of the tourism sector, since it acts as a catalyst for creating favourable conditions for the development of local production;
- It is necessary to show the tools and to teach the local population how to use them correctly for a careful attitude to the natural and socio-economic resources.

At the same time, the participation of residents provides more effective planning and decision-making, and allow with their support, to avoid mistakes that could later provoke conflicts with the local population itself. Participation of the local population increases the responsibility of other participants in process, and officials will contribute to the formation of unified approach concerning the spectrum existing of problems.

Moreover, as practice shows, it is quite difficult to ensure a stable relationship between the benefits received by local communities from ecotourism and the strengthening of the protection of the natural territories. Moreover, there are certain technical and organizational problems associated with this issue.

The cost of the project increases due to the increase in the number of necessary personnel (managers, coordinators) from the local population. Benefits do not always reach the target groups for which they were planned. Detailed information for residents about the tourism development project increases their distrust of the project, but on the other hand, if work with the local population is performed at a high level, this will certainly give a high result.

When developing ecological tourism projects, it is also necessary to consider the fact that the needs and desires of residents in different communities (and even in the same one) may differ greatly (Brandon 1993).

Option 3. Participation of local people in the project of ecotourism development can be economic and/or political. The meaning of the economic is, first, to participate in distribution of income derived from the ecotourism business (Khrabovchenko 2007; Panov 2000).

Economic participation of local communities can be in the following forms:

- self-employment, carried out individually or in groups/communities;
- joint ventures with third-part professional tourist companies;
- hired work (employees of travel companies);
- provision of local services to tour operators (supply of food, construction materials, handicraft products, etc.);
- deduction of income from the ecotourism business to protected area as compensation for environmental management or for the operation of utility facilities;
- payment for the transfer of rights to use local facilities (buildings, vehicles, etc.) to tourist enterprises (rent, etc.).

All the above provides the local population with additional income or other benefits and advantages. For example, according to research by the US experts, when organizing eco-tours by American tour operators, citizens of the host country are provided with the following types of work: guides and translators—20% of those employed in the field of ecotourism, drivers—8%, porters—8%, cooks—6%, and also hired managers and tour operators—8%. In general, local participation is about 50%. Nevertheless, of course, this does not mean full participation in decision-making, i.e. political participation.

Unfortunately, as world experience shows, most ecotourism development projects operate within the framework of commitment to the principle of profit sharing. Fundamental decisions concerning the project nature, issues of employment, general type of development made without considering the opinion of the local population.

The first step to involve the local population in the project should be to collect information. Reliable information about the local population and their preferences is the basis for the development of adequate projects and plans for the development of ecotourism.

At the information gathering stage, it is usually possible to identify local leaders, the nature and mission of local organizations, the main priorities of local communities, as well as what ideas, expectations and doubts residents associate with the development of ecotourism. It is important to clarify which types of activities and which local products can be most acceptable in tourism and will bring the greatest benefit to the entire local community as a whole? For example, it can be national games and traditions, Kazakh national cuisine, arts and crafts, etc.

New ideas obtained in the process of collecting information, as well as all other available information (the results of marketing research, government plans for territory development, etc.) should be submitted for discussion to local communities. For this, it is necessary to hold various communication activities, spread information and involve representatives of various groups in the discussion. The most important component of this phase is education (trainings), which both local organizations at various levels and plans, and residents may need.

Information gathering and discussion activities should be carried out to provide local people and organizations with an opportunity to feel that they are directly related to the project. Moreover, in parallel with the collection of information, consultations of the local population are usually held.

The population should be aware that their income is linked to the need to protect the resources that form the basis of their tourism business and, consequently, their well-being. Otherwise, the development of ecotourism will not lead to increased protection of nature.

To create positive incentives for nature conservation, it is necessary to strengthen the relationship between environmental goals and the potential benefits of ecotourism. These advantages should be significant for the majority of the population. For example, from an environmental point of view, it is more effective to provide part-time employment as a guide for many potential poachers, rather than hiring one or two full-time employees. Benefits provided by ecotourism must be flexible to meet the interests of the most diverse groups of local people.

For local population, the relationship between protection of resources and obtaining of known benefits by them, i.e. profits. Ecotourism will become an incentive for nature conservation and catalyst for local development if the development benefits are enjoyed by significant part of local population.

To ensure participation in ecotourism development of the local population, it is possible to use two approaches.

The first (individual approach) provides for the involvement of individual third-party expert-specialists in the project. The second one is based on creation (attraction) for participation in ecotourism development of local organizations.

In the first case, the result in form of changes on local ideas, technologies and the development of new activity types is achieved, as a rule, faster. However, in the end, this results in less sustainable results than participation through an organization. The organization can be a flexible tool for attracting both individuals and collective project participants, acting simultaneously as a link between residents and external structures (administrations of various levels, non-profit organizations, tour operators and travel agencies). In addition, to strengthen such organizations, projects may include both training of their members and direct leadership.

A combined approach is possible when experts work directly with local organizations. If a possible implementation of this form of work appears to be most optimal for achieving the objectives of the project in both the short and long term.

It is much easier for residents to participate if they can act through any organization they know, especially in cases where a huge area needs to be covered.

It should be noted that there is no universal model for attracting local people to ecotourism. In almost every case, there are restrictions on this issue (Ecotourism development prospects in Kazakhstan 2010).

They may be due to local specifics, limited funding and periods for organizing ecotourists reception, difficulties in reaching consensus between individual public groups, etc. Moreover, sometimes-local authorities can restrain wide public participation in decision-making.

One of the conditions for success is the existence of mutually beneficial cooperation between NPOs, residents, public authorities and private sector. At the same time, optimal level of local participation is determined by specificity and scale of a certain project affecting the interests of residents on individual settlements or number of communities (Baikanova 2012). Choice of the optimal approach depends on specific socio-economic conditions, population mentality and region resources. If the projects is flexible enough and open for discussion with the local population, then the probability of successful planning on ecotourism development is almost high.

Implementation of plans for ecological tourism development in protected areas (for example, SNPP "Ile-Alatau") may initiate the cooperation of park administration with residents/local communities of settlements located along the perimeter of protected areas. In this case, there are villages of Turgen, village of Tau-Turgen, Esik city, village of Ryskulovo, village of Orman, village of Beskainar (mountain gardener), village of Kokshoky, rotational settlement at the Big Almaty Lake, Izvestkovyi settlement, residential areas based on the cottage settlements, etc.

Today, due to the active development of industrial technologies and conveyor production, standardized products of the craft industry are becoming uncompetitive. Thanks to rural tourism, you can revive handicraft goods in the form of souvenirs, elements of national clothes, promote the tourist image of the area, arrange demonstration workshops, and use traditional handicrafts in the arrangement of rooms in guest houses and much more.

Algorithm for developing concepts for ecotourism products at the rural level:

1. Definition of parameters for the development of the tourism sector in this area;
2. Determining the specifics of the tourism potential and on its basis the development of a comprehensive concept of the total tourism product, considering their tourism resources and offers of neighbours in the tourism cluster;
3. If a strategic decision on tourism was made at the local level development, and then ecotourism, as a direction of the tourism industry, is a prerequisite for rural areas due to the peculiarities of their potential;
4. Development of models for ecotourism development in a specific rural area and choice of the most appropriate development option.

Due to the relatively limited scale of rural eco-tourism in comparison with other types of tourism, its socio-economic importance is particularly high at the local and regional levels.

The development of ecotourism in rural areas generates so-called non-monetary benefits that cannot always be estimated economically or that are only indirectly expressed by economic categories. These include, for example:

- the creation of new jobs;
- environmental education;
- raising the cultural level of the population;
- development of the network infrastructure;
- economic growth in backward areas.

In cases where rural tourism is not initially a highly profitable enterprise at the local level, some States often encourage its development through subsidies, taxes, and other benefits, since indirect advantages at the regional and national levels make it preferable to develop it.

Direct or indirect rural tourism also has a strong influence on the development of other sectors of the economy—primarily agriculture, transport, construction materials, individual entrepreneurship and the development of crafts.

Tourism revenues allow for more efficient use of resources and modern technologies. This gives a significant boost to the development of the economy and the well-being of the local population, especially in areas remote from major industrial centres, with a traditionally low level of economic development.

References

Aydpakelov NS (2017) Tourism of Kazakhstan 2012–2016. In: Statistical collection in the Kazakh and Russian languages, Astana, p 86

Baikanova DE (2012) Development of ecotourism in the Republic of Kazakhstan. Vestnik №1. Tourism. Available at http://tourlib.net/statti_tourism/bajkanova2.htm. Accessed 21 Mar 2018

Blinova TV, Kutenkov RP, Rubtsova VN (1999) Social sustainability of the rural community. Soc Res 8:35–38

Brandon K (1993) Ecotourism: guide for planners and managers. North Bennington, p 175

Centre for tourism development of the Sverdlovsk region (2011) Rural and ecological tourism as a factor of sustainable development of the territory. Guidelines, Yekaterinburg, p 32

Diamantis D (2010) The concept of ecotourism: evolution and trends. Curr Issue Tour 2(2–3):93–122

Drozdov AV (2010) Ecological and tourist potential: definition, components, assessment. Actual Probl Tour 4:12–23

Foreigners arriving in Kazakhstan will be able to register at hotels (2014) Available at http://news.mail.ru/inworld/Kazakhstan/society/18330376/. Accessed 10 Aug 2016

Final report on the development of systemic plan for tourism in the Republic of Kazakhstan (2012) Astana, p 930

Global Consulting in Tourism—World Travel Monitor. Official website of international research company in the field of tourism. IPK International. Available at http://www.ipkinternational.com. Accessed 25 June 2020

Kazakhstan Tourist Association. Official website of Kazakhstan Tourist Association. Available at http://www.kaztour-association.com. Accessed 17 May 2020

Kazakov NP (2009) Theory and methodology of development on ecological entrepreneurship in the recreational sphere. Abstract of dissertation on the degree of economic sciences doctor, Saint-Petersburg, p 42

Khrabovchenko VV (2007) Ecological tourism. Finance and Statistics, Moscow, p 208

Map of Kazakhstan. Website of stock maps images on bestmaps site. Available at https://bestmaps.ru/strana/kazahstan. Accessed 10 Nov 2020

Minnaert L, Maitland R, Millera G (2011) What is social tourism? Curr Issues Tour 14(5):403. Special issue: social tourism: perspectives and potential

Nikonorov SM (2013) Ecotourism development: a regional model. On the example of the Volga federal district. Institute for Sustainable Development on Public Chamber of the Russian Federation/Centre for Environmental Policy of Russia, Moscow, p 120

Panov IN (2000) Ecotourism and its role in the sustainable development of territories, Series 6. Geography, № 6. Vestnik of Moscow State University, Moscow, pp 117–127

Prospects for the ecotourism development in Kazakhstan (2010) J World Monit 2(23):215–230. EUROBAK (European Business Association of Kazakhstan)
Rakhmanovskie spring photo. Official website of Google—Rakhman spring. Available at https://www.google.kz/search?q=рахмановские+ключи+фото&rlz. Accessed 10 Nov 2020
Samoilenko AA (2006) Geography of tourism. Study guide. "Phoenix", Rostov-on-Don, p 368
Schwab K, Martin C, Samans R, Moavenzadeh Zh, Drzeniek-Hanouz M (2017) The travel & tourism competitiveness report 2017. World Economic Forum, Geneva, p 387
Shaykenova RR, Mamutova KR (2010) Current state of ecological tourism in Kazakhstan. Bull Turan Univ 45(1):87–90
Statistical yearbook of Kazakhstan for 2010, 2011, 2012 years (2012) Astana, p 540
Suvorov ZV (2017) Ecological path as a means of shaping the ecological culture of schoolchildren. Educ J "Pedagogy" 27
Turner R (2018) Travel and tourism: economic impact 2018, Kazakhstan. WTO, London, pp 1–24
Vorotnikova IL (2012) Agro tourism: experience, problems, solutions. In: Materials of the international scientific-practical conference. Federal State-Funded Educational Institution of Higher Professional Education "Saratov SAU", p 184
Yerdavletov SR, Aktymbaeva AS, Iskakova KA et al (2013) Report on research work: development of the concept on regional development and territorial organization in internal tourism of the Republic of Kazakhstan (interim). Almaty, p 181
Zaytseva OO (2013) Rural areas as an object of management: the concept, functions, typology. Fundam Res 6(part 2):416–420

Conclusion

7

The following results were obtained during studies conducted by the authors:

1. Essence and content of ecological tourism, factors of its development are determined. Ecological tourism is a type of tourism that contributes to the preservation of environment, based on tourist demand, developing in a cultural context.
2. The role of ecological tourism in the tourism system has been established, and mechanisms have been identified to ensure the self-preservation and improvement of the tourism system and main objectives of ecological tourism development in the Republic of Kazakhstan.
3. Classification of ecological tourism has been developed, its types have been identified depending on the specifics of tourism activities and on their attitude to environmental aspects of recreation.
4. The natural and recreational potential of the Republic of Kazakhstan allows to fully develop ecological tourism in the country. In general, the nature of the republic is diverse, there are many the natural complexes that are well-preserved in there the natural state, which is important for ecological tourism development.
5. Socio-economic tourism potential of Kazakhstan serves as a prerequisite for the organization of ecological tourism. On the territory of the republic there is a significant number of monuments reflecting the history of the country from ancient times to the present day. All of them are of great interest to tourists, acting as environmental tourism facilities.
6. According to the results of studies, authors zoned the territory of the Republic of Kazakhstan according to the combined tourist and recreational potential. Zoning for tourism and recreational purposes requires consideration of complex factors —the natural, economic, physiological, environmental, social, architectural, artistic and many others. Recreational zoning methods are based on various parameters, both the scale of zoning and the determination of tourist and recreational potential of territory for possible study on purpose of economic development.

K. Iskakova et al., *Ecological Tourism in the Republic of Kazakhstan*,
https://doi.org/10.1007/978-3-030-77462-2_7

7. Assessment and zoning of tourism and resource potential of ecological tourism in the regions of Kazakhstan made it possible to identify that in West Kazakhstan areas there are more than 60 most attractive and studied objects, from 41 to 60, from 21 to 40, and less than 20 objects.
8. Analysis of the natural prerequisites' assessment on development of tourism and recreation in North Kazakhstan and landscape classification by the degree of attractiveness made it possible to allocate territories with the most favourable recreational resources—these are Korgalzhyn, Shchuchinsk, Yereymentau, Sandyktau, Zerendy districts of Akmola region. For the development of certain types of tourism, the territories of Kostanay and Pavlodar regions are promising ones. Territory of the North Kazakhstan region is the least attractive for mass tourism.
9. Assessment results of tourism and recreational potential of Central Kazakhstan showed that the Karkaraly district and the cities of Karaganda, Temirtau have favourable economic and geographical conditions and a high level of recreational potential, which have the best prospects for tourism development.
   The Aktogay and Ulytau districts have relatively favourable conditions and an average level of tourist potential. Aktogay district and the city of Balkhash are allocated due to the location around Lake Balkhash. Ulytau district and the city of Zhezkazgan, due to the presence of significant cultural and historical prerequisites, attractive the natural landscapes and a fairly high level of infrastructure development, belong to the territories relatively favourable for ecological tourism development.
   Areas, which characterized by insufficiently favourable conditions and low tourist potential are Osakarovsk, Bukharzhyrau, Abay and Shet districts and with low recreational potential—Nura and Zhanaarka.
10. South Kazakhstan region (Turkestan region) has a variety of tourist and recreational resources and can develop all types of tourism. When assessing the total tourism and recreational potential of the region, it turned out that Almaty region has a high potential; the average potential is characteristic of South Kazakhstan region; the below average potential of Zhambyl region; and Kyzylorda region has low potential.
    Authors of a monograph believe that although the Kyzylorda and Zhambyl regions have low scores, for ecological tourism development, these regions have all the necessary resources, especially within specially protected the natural areas.
11. According to the assessment results, East Kazakhstan has the highest potential than other regions of Kazakhstan. The districts of Katon-Karagay, Urzhar received the maximum number of points—10, all other districts of the East Kazakhstan region have an average potential (7 points), and the Beskaragay region has below average potential (6 points). In general, northern and eastern parts of a region are favourable zones for the development of recreational activities.
12. Analysis of tourism and recreational resources of the Republic of Kazakhstan made it possible to conclude that to attract foreign tourists there is a minimum

infrastructure, but there are necessary resources for the ecological tourism development. Nature reserves, protected areas and rural resources are of particular importance for the ecological tourism development. Betterment of ecological tourism based on the natural resources of Kazakhstan carry with it a positive socio-economic effect for the local population and country as a whole, which in turn increases the opportunity to rise competitiveness of domestic tourist facilities and destinations in tourism market both in Central Asia and far abroad.

13. During the analysis of ecological tourism current state in the country, the authors identified the need to create and modernize infrastructure of tourism industry—tourist accommodation, transport and logistics, service related to excursion activities and production of souvenir products.
14. Results obtained above can be used in further detailed study of problems on ecological tourism development. Cartographic material and proposed events will allow developing optimal variations for solving the problems on integrated use of the natural and recreational resources, and may also be in demand by production, scientific and other organizations that aim to develop ecotourism and preserve the natural wealth and historical-archaeological sites for the future generation.

Further improvement and development of theoretical, methodological and applied provisions set forth in monograph should be aimed at the issues of a fundamental enhance on quality of use concerning the natural and recreational resources to ensure a more complete solution to the betterment of tourism industry in Kazakhstan and increase the flow of tourists to the country.

The manufacturer's authorised representative in the EU is Springer Nature Customer Service Centre GmbH, Europaplatz 3, 69115 Heidelberg, Germany. If you have any concerns regarding our products, please contact ProductSafety@springernature.com

Printed and bound by CPI Group (UK) Ltd, Croydon, CR0 4YY
30/09/2026
02225691-0005